Table of Contents

Laboratory Exercise 1: Diffusion
(adapted from Warren Dolphin's *Biological Investigations, Sixth Edition.*)

Supplies
Equipment
Balances

Materials
Bottle of potassium permanganate crystals
Bottle of carmine powder
For each student group:

> Dialysis tubing cut in 15 cm sections
> 4 100-ml beakers
> 2 10-ml pipettes
> 50 ml of .25% starch in 1% Na_2SO_4
> 200 ml of 1% albumin in 3% NaCl
> Medicine droppers with:
>> 1 M $AgNO_3$
>> 2% $BaCl_2$
>> I_2KI
>> Bradford Reagent
>
> 1 test tube rack with 16 test tubes
> 2 Petri dishes
> 2 15-cm (6-inch) rulers
> 2 dissecting probes

Objectives:
1) To determine the type of molecules which can move across a selectively permeable membrane via simple diffusion;
2) To visualize Brownian motion
3) To determine the relationship between distance traveled and the rate of diffusion.

✠Characteristics of a selectively permeable membrane.
The maintenance of a constant internal environment in a cell or organism is called **homeostasis**. In a constant environment, enzymes and other cellular systems are able to function at optimum efficiency. One component of a cell's homeostatic mechanisms is the ability to exchange materials

selectively with the environment. Ions and organic compounds, such as sugars, amino acids, and nucleotides, must enter a cell, whereas waste products must leave a cell. Regardless of the direction of movement, the common interface for these processes is the plasma membrane. The cell walls of plants and bacteria offer little, if any resistance to the exchange of molecules.

The plasma membrane is a mobile mosaic of lipids and proteins (Figure 1.1). Materials cross this outer cell boundary by several processes. Large particles are engulfed into a membranous organelle, forming a vesicle or vacuole that can pass into or out of the cell. Some small molecules diffuse through the spaces between lipid molecules in the membrane. Others bind with proteins in the membrane and are transported into or out of the cell.

To understand cellular transport, you should recognize that atoms, ions, and molecules in solution are in constant motion, continuously colliding with one another because of their **kinetic energy**. As the temperature of any phase is raised, the speed of movement increases so that molecules collide more frequently with greater force. A directly observable consequence of this constant motion is **Brownian movement**, an erratic, vibratory movement of small molecules in aqueous suspension caused by collisions of water molecules with the particles.

Diffusion also results from the kinetic energy of molecules. For example, when a few crystals of a soluble substance are added to water, molecules break away from the crystal surface and enter solution, some traveling to the remotest regions of the solution. This process continues until the substance is equally distributed throughout the solvent. To generalize this example, in any localized region of high concentration, the movement of molecules is, on the average, away from the region of highest concentration toward the region of lowest concentration. The gradual difference in concentration over the distance between high and low regions is called the **concentration gradient**.

Figure 1.1. The fluid mosaic model of a cell membrane.

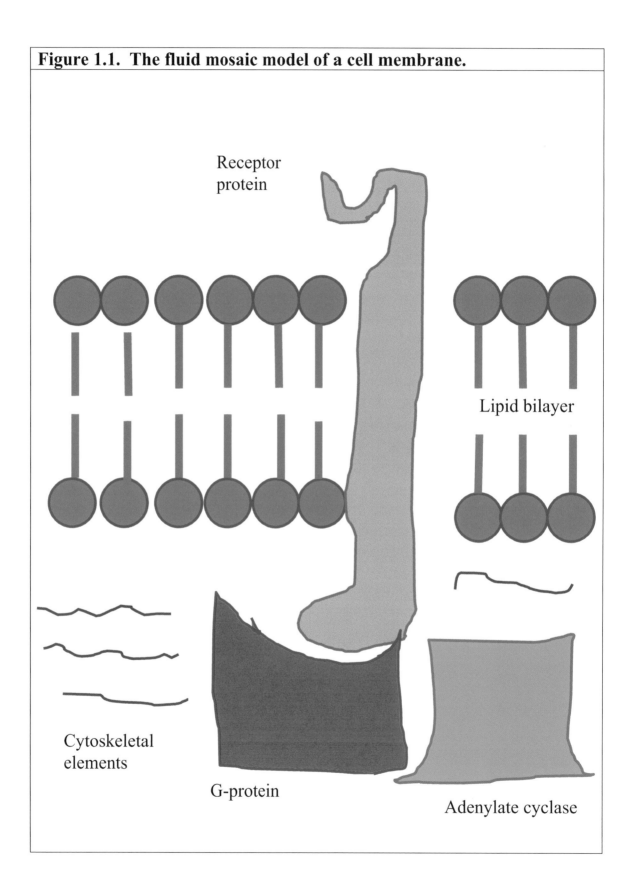

Receptor
protein

Lipid bilayer

Cytoskeletal
elements

G-protein

Adenylate cyclase

The steeper the concentration gradient, the more rapid the rate of diffusion. The rate of diffusion is also directly proportional to temperature and inversely proportional to the molecular weight of the substance involved. (All molecules move rapidly at high temperatures, but larger molecules move more slowly than small molecules at the same temperature.)

Substances diffuse into and out of cells by passing through the spaces between membrane molecules or dissolving in the lipid or protein portions of the membrane. However, due to size or charge, some substances cannot pass through membranes. Membranes that block or otherwise slow passage of certain substances are described as being **differentially permeable**. Differential permeability accounts for the phenomenon of **osmosis**, or the diffusion of **water** through a membrane.

Water will diffuse through a membrane when there is a concentration gradient between the interior and exterior of the cell. Eventually, an equilibrium would be reached when the flow of water into the cell, due to concentration differences, balances the flow out of the cell, caused by pressure differences. The pressure at equilibrium is called the **osmotic pressure** of the solution. Two examples which demonstrate the importance of osmotic pressure include:
1) The return of water lost at the arteriole end from a capillary, because of hydrostatic pressure, back into the venule end of a capillary, so that the net loss of fluid from the blood entering a capillary bed is less than 2%;
2) The opening and closing of flowers and leaves, as in the Sensitive Plant, *Mimosa pudica*.

Since all cells contain molecules in solution that cannot pass through the membrane, osmosis always occurs when cells are placed in dilute aqueous solutions. In bacteria and plants, the cell wall prevents the cell from bursting by providing a rigid casing that helps regulate osmotic pressure within the cell. In animals, an osmoregulatory organ is found, such as the kidney, which adjusts the concentration of substances in the body fluids that bathe the cells.

Many ions and organic molecules important to cell metabolism are taken into cells by specific transport proteins found in the cell membranes.

Facilitated diffusion occurs when such a protein simply serves as a binding and entry port for the substrate. In essence, the protein is a pipeline for a specific substance. The direction of flow is always from a region of high concentration to one of low concentration, but gradients are maintained because many molecules, upon entering the cell, are metabolically converted to other types of molecules.

For many other materials, favorable diffusion gradients do not exist. For example, in mammalian neurons, sodium ions are found at high concentrations outside the cells, yet the net movement is from inside to outside the cell. At the same time, potassium ions are found high concentrations inside the cells, yet the net movement is from outside to inside the cell. (Table 1.1.) For such materials, cellular energy must be expended to transport molecules across the cell membrane. **Active transport** occurs when proteins in the cell membrane bind with the substrate and with a source of energy to drive the "pumping" of a material into or out of a cell.

Table 1.1. Comparison of ion concentrations within mammalian neurons, in mM (Adapted from Campbell, 1996)

	Na^+ (sodium)	K^+ (Potassium)	Cl^- (Chloride)	A- (Anions, including proteins and amino acids)
Inside Neuron	15	140	10	100
Outside neuron	150	5	120	~0

Procedure:

Diffusion and osmosis can be demonstrated simultaneously in one setup. Dialysis tubing is an artificial membrane material with pore sizes that allow small molecules to pass through it but not large molecules.

Water, NaCl, and Na_2SO_4 have molecule weights of 18, 58.5, and 146, respectively. Starch and proteins have molecular weights greater than 100,000. If dialysis tubing is a differentially permeable membrane, which molecules would you hypothesize can pass through the membrane?

1) Obtain a 30 cm section of dialysis tubing that has been soaked in distilled water. Tie or fold and clip one end of the tubing to form a leakproof bag. Half fill the bag with a solution of 1% protein (albumin) dissolved in 3% NaCl. Also add a 3 ml sample of the same solution into each of four test tubes labeled "**Inside Solution Start**".
2) Now tie the bag closed with a leakproof seal or knot. Wash the bag with distilled water, blot it on a paper towel, weigh it to the nearest 0.1 gram, and record the weight in Table 1.2.
3) Place the bag in a 250 ml beaker containing a solution of 0.25% soluble starch dissolved in 1% Na_2SO_4. Place 3 ml samples of the fluid from the beaker into each of four test tubes labeled "**Outside Solution Start**." The starting conditions are summarized in Figure 1.2. This experiment will run for approximately 1 hour. Go on to the other experiments while this experiment runs in the background.
4) At 15-minute intervals, swirl the beaker containing the bag or place the beaker on a slowly turning magnetic stirrer.

Figure 1.2. Starting conditions for osmosis and diffusion experiment, showing the composition of solutions inside and outside the dialysis bag.

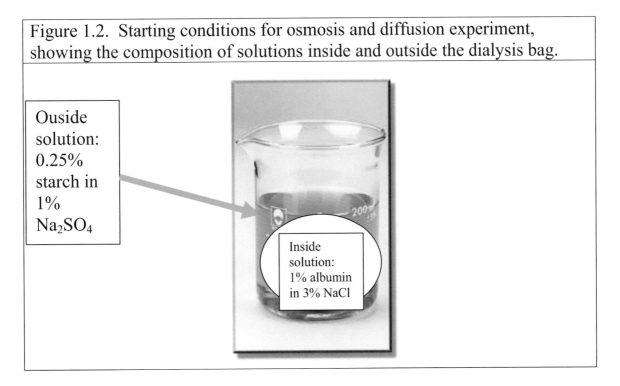

Ouside solution: 0.25% starch in 1% Na_2SO_4

Inside solution: 1% albumin in 3% NaCl

Analysis:
1) After 1 hour, take four 3-ml samples from the beaker and place them in four test tubes labeled "**Outside solution end**". Now remove the bag,

rinse it with distilled water, blot it on a paper towel, and weigh it to the nearest 0.1 gram. Record the weight in Table 1.2.

2) Empty the contents into a beaker, take four 3-ml samples and place them in four test tubes labeled "**Inside solution end**".
3) Assay the inside and outside samples from the start and end for the presence of the compounds added at the beginning of the experiment. Record the results of your analysis in Table 1.2, using the plus and minus symbols to indicate the presence or absence of material both before and after incubation. The following are specific, easy-to-perform indicator tests:

Test for Chloride ion: Add a few drops of 1 M $AgNO_3$ to one inside and one outside tube for both start and end samples. A milky white precipitate of AgCl indicates the presence of Cl-.

Test for Sulfate Ion: Add a few drops of 2% $BaCl_2$ solution to one inside and one outside tube for both start and end samples. If SO_4^- is present, a white precipitate of $BaSO_4$ will form.

Test for Protein: Add 1 ml of Bradford Reagent to the test tube. If protein is present, the solution will turn a bright blue color.

Test for Starch: Add a few drops of I_2KI to each remaining tube. If a blue color appears before mixing, it indicates the presence of starch. If no color develops, add a few crystals of KI without mixing, then add I_2 crystals. IF a blue color develops as the iodine dissolves but then disappears, this is still a positive test for starch.

Which set of test tubes served a control in this experiment?

Describe which ions were able to move through the dialysis membrane. Which direction did they move in relation to their concentration gradient? What are the molecular weights of these ions.

✠ Did starch and protein move through the dialysis membrane? What are their typical weights?

✠ What evidence do you have that water moved through the dialysis membrane?

Table 1.2. Results of osmosis/diffusion experiment with dialysis tubing.				
	Outside solution		Inside solution	
	Start	End	Start	End
NaCl				
Na$_2$SO$_4$				
Protein				
Starch				
H$_2$O (weight)	XXXXXXX	XXXXXXX		

✠Brownian movement

The vibratory movement exhibited by small particles in suspension in a fluid was first observed in 1827 by Robert Brown, a Scottish botanist. Brown erroneously concluded that living activity caused this motion, but scientists now know that Brownian movement results from the constant collision of water molecules with particles. Small particles 10 micrometers or less in size are noticeably displaced by the collision, whereas larger particles are not. (For comparison, red blood cells are 7 micrometers in diameter.)

To illustrate Brownian movement, place a drop of water in the center of a microscope slide. Dip a dissecting needle into a bottle of carmine stain and then roll the needle tip in the water drop. Add a coverslip so that the drop of water is sandwiched between the slide and the coverslip.

Briefly record your impressions of the movement of carmine stain particles.

�֎Diffusion Rate as a Function of Distance

Procedure:

1. Add 15 ml of water into a Petri dish. Allow the water to settle, and do not move the Petri dish.
2. Place a 6-inch (15-cm) ruler over the top of the Petri dish.
3. Place one potassium permanganate $KMnO_4$ crystal into the middle of the Petri dish.
4. Measure the diameter of the cloud every 15 seconds for 3 minutes, and record your results in Table 1.3.
5. Plot the diameter of the cloud as a function of time on one graph, labeled as Figure 1.3.
6. Plot the velocity of diffusion, as a function of cloud diameter on the second graph, labeled as Figure 1.4.

✖ As the diffusion distance increases, what happens to the diffusion velocity?

Table 1.3. Diffusion of potassium permanganate

Elapsed time (sec)	Diameter of the cloud (cm)	Net velocity of diffusion (cm/second) (=diameter/elapsed time)
0	0	------------
15		
30		
45		
60		
75		
90		
105		
120		
135		
150		
165		
180		

Figure 1.3. Cloud diameter (cm) as a function of time (seconds). (Fill in the appropriate units on the y-axis)

Figure 1.4. Diffusion velocity (cm/second) as a function of cloud diameter (cm). (Fill in the appropriate units on the y-axis)

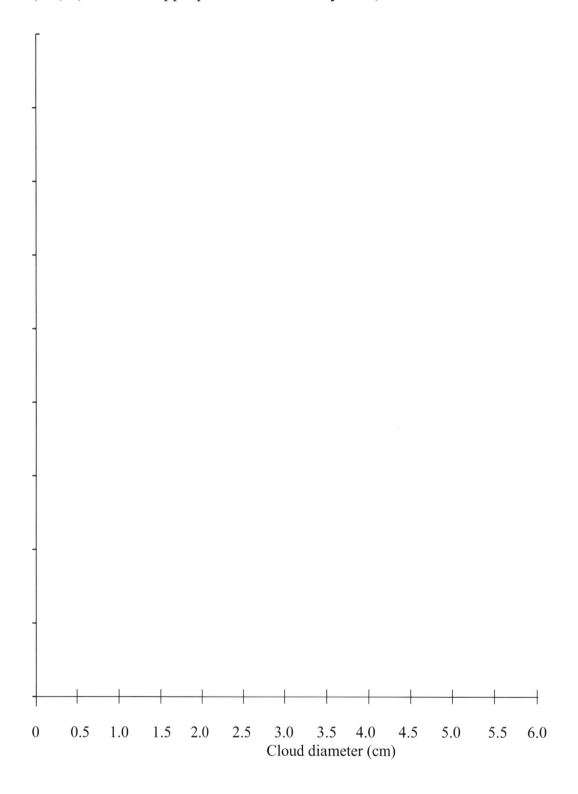

References Cited:

Campbell, N. 1996. Biology, edition 4. Benjamin Cummings.

Dolphin, W.D. 2002. Biological Investigations: Form, Function, Diversity, and Process, edition 6. McGraw-Hill, Dubuque, Iowa.

Exercise 2: Enzymes
(Adapted from Warren Dolphin's Biological Investigations, edition 6, McGraw-Hill)

Prelab Exercise:
A pre-lab exercise is accessible via, which describes peroxidase activity as a function of temperature:
http://bioweb.wku.edu/courses/Biol114/enzyme/enzyme1.asp

Equipment
Constant temperature water baths or large trays to serve as water baths
Spectrophotometers at 500 nm

Materials
Blender or mortar and pestle
Fresh white turnip, horseradish root, or potato
Tissues and markers
Optical surface tubes for the spectrophotometer
15 ml test tubes and rack
50 ml beakers
5 ml pipettes graduated in 0.1 ml units with suction devices or automatic pipetters
Thermometer

Solutions
10 mM H_2O_2
25 mM guaiacol (Sigma Chemical Co.)
Citrate-phosphate buffers at pHs 3, 5, 7, 9

✠Prelab Preparation
Before doing this lab, you should read the introduction and sections of the lab topic that have been scheduled by the instructor.

Do the Virtual Enzyme lab, which describes peroxidase activity as a function of temperature. This virtual lab is accessible via:
http://bioweb.wku.edu/courses/Biol114/enzyme/enzyme1.asp

You should be able to describe, in your own words, the following concepts:
1) Structure of an enzyme
2) Effect of pH on enzyme structure
3) Effect of temperature on enzyme structure
4) Operation of a spectrophotometer

Objectives

1) To perform a quantitative assay of the activity of an enzyme in a tissue extract using a spectrophotometer
2) To organize the data as concise tables and graphs for inclusion in a lab report describing the properties of the enzyme peroxidase
3) To test the following null hypotheses:
    a. The temperature of the solution does not influence the activity of an enzyme.
    b. The pH of the solution does not influence the activity of an enzyme.

Introduction

The thousands of chemical reactions occurring in a cell each minute are not random events but are highly controlled by biological catalysts called **enzymes**. Like all catalysts, enzymes catalyze reactions by lowering the **activation energy** of a reaction, the amount of energy necessary to trigger a reaction (Figure 2.1).

Most enzymes are proteins with individual shapes determined by their unique amino acid sequences. Graphic representations of enzymes usually omit the individual amino acids in the sequence, but they will show the alpha sheets, beta helices, and the active site, as in the representation of peroxidase shown in Figure 2.2. Since these sequences are spelled out by specific genes, the chemical activities of a cell are under genetic control. The shape of an enzyme, especially in its **active site**, determines its catalytic effects. The active site of each type of enzyme will bind only with certain kinds of molecules -- for example, some enzymes bind with glucose but not with ribose because the former is a six-carbon sugar while the latter has only five carbons.

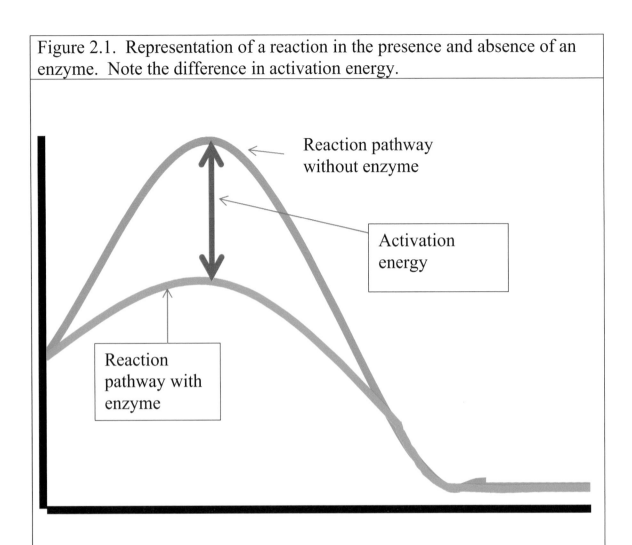

Figure 2.1. Representation of a reaction in the presence and absence of an enzyme. Note the difference in activation energy.

Reaction pathway without enzyme

Activation energy

Reaction pathway with enzyme

A molecule that binds with an enzyme and undergoes chemical modification is called the **substrate** of that enzyme. Often metallic ions, such as iron (Fe^{+++}), magnesium (Mg^{++}), calcium (Ca^{++}), or manganese (Mn^{++}), aid in the binding process, as do vitamins or other small molecules called **co-factors** or coenzymes.

The binding between enzyme and substrate consists of weak, noncovalent chemical bonds, forming an **enzyme-substrate complex** that exists for only a few milliseconds. During this instant, the covalent bonds of the substrate either come under stress or are oriented in such a manner that they can be attacked by other molecules, for example, by water in a hydrolysis reaction.

The result is a chemical change in the substrate that converts it to a new type of molecule called the **product** of the reaction. The product leaves the enzyme's active site and is used by the cell. The enzyme is unchanged by the reaction and will enter the catalytic cycle again, provided other substrate molecules are available (Figure 2.3).

Figure 2.2. Graphic representation of an enzyme with a substrate molecule in the active site.

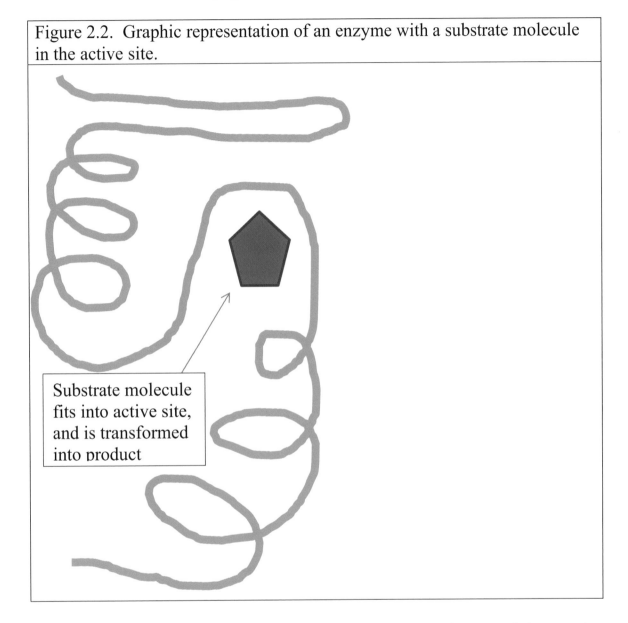

Substrate molecule fits into active site, and is transformed into product

Individual enzyme molecules may enter the catalytic cycle several thousand times per second; thus, a small amount of enzyme can convert large quantities of substrate to product. Eventually enzymes wear out; they break apart and lose their catalytic capacity. Cellular proteinases degrade inactive enzymes to amino acids, which are recycled by the cell to make other structural and functional proteins.

The amount of a particular enzyme found in a cell is determined by the *balance* between the processes that *degrade* the enzyme and those that *synthesize* it. When no enzyme is present, the chemical reaction catalyzed by the enzyme does not occur at an appreciable rate. Conversely, if enzyme concentration increases, the rate of the catalytic reaction associated with that enzyme will also increase.

Figure 2.3. Catalytic cycle.

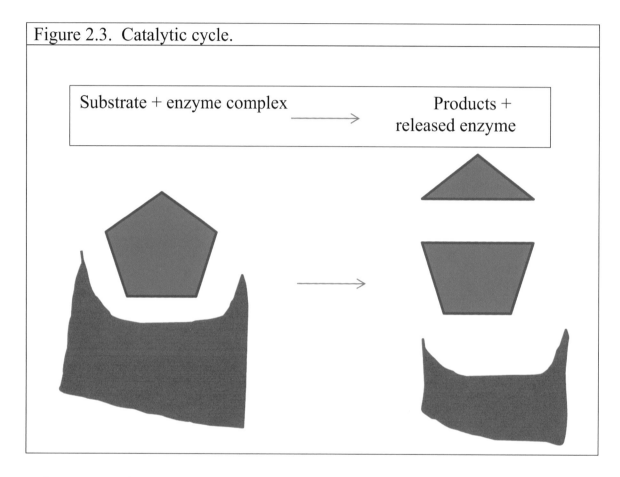

The pH or salt concentration of a solution affect the shape of enzymes by altering the distribution of + and - charges in the enzyme molecules which, in turn, alters their substrate-binding efficiency. Temperature, within the physical limits of $0°$ and $40°C$, affects the frequency with which the enzyme and its substrates collide and, hence, also affects binding. All factors that influence binding obviously affect the rate of enzyme-catalyzed reactions. Some of these factors will be investigated during this laboratory.

Peroxidase

During this lab, you will study an enzyme called **peroxidase**. It is a large protein containing several hundred amino acids and has an iron ion located at its active site. Peroxidase makes an ideal experimental material because it is easily prepared and assayed. Turnips, horseradish roots, and potatoes are rich sources of this enzyme.

The normal function of peroxidase is to convert toxic hydrogen peroxide (H_2O_2), which can be produced in certain metabolic reactions, into harmless water (H_2O) and oxygen (O_2):

$$2\ H_2O_2 \xrightarrow{\text{Peroxidase}} 2\ H_2O + 2\ O$$

The oxygen often reacts with other compounds in the cell to form secondary products.

The peroxidase reaction can be measured by following the formation of oxygen. The amount of oxygen present after the reaction can then be measured, indirectly, by the degree to which a dye turns color in the presence of oxygen.

The dye we will use in this experiment is called **guaiacol**, and it turns brown when oxidized. The entire peroxidase reaction, including the measure of active oxygen through guaiacol, is as follows:

$$2\ H_2O_2 \xrightarrow{\text{Peroxidase}} 2\ H_2O + 2\ O$$

$$\underset{\text{(colorless)}}{O + guaiacol} \longrightarrow \underset{\text{(brown)}}{oxidized\ guaiacol}$$

To quantitatively measure the amount of brown color in the solution in which the reaction is occurring, the enzyme, substrate, and dye can be mixed In a tube and immediately placed in a spectrophotometer. As color accumulates, the absorbance at 500 nm will increase.

Preparing an Extract Containing Catecholase

These steps will be done by the instructor before class to save time:
1) Weigh 1 to 10 grams of peeled turnip, horseradish root, or potato tissue on a double- or triple-beam balance;
2) Homogenize the tissue by adding it to 100 ml of cold (4°C) 0.1 M phosphate buffer at pH 7. Grind the mixture in a cold mortar and pestle

with sand or blend it for 15 seconds at high speed in a cold blender. The extract will keep overnight in a refrigerator.

Factors affecting Enzyme Activity

Temperature Effects

To determine the effects of temperature on catecholase activity, you will repeat the enzyme assay in water baths, at four temperatures:
1) In an ice bath, approximately 4°C
2) At room temperature (about 23°C)
3) At 32°C
4) At 48°C

State the null (H_0) hypothesis that relates change in enzyme activity to the temperature of the solutions used:

H_0:

State an alternative hypothesis:

H_a:

To test your null hypothesis, you should use the following directions to set up the reactions and conduct the experiment:

If constant temperature baths are not available, improvise with plastic containers, adding hot and cold water to adjust the temperature. Number nine test tubes in sequence 1 through 9. Refer to Table 2.1 for the volumes of reagents to be added to each tube.

Preincubate all the solutions at the appropriate temperature for at least 15 minutes before mixing. After reaching temperature equilibrium and adjusting the spectrophotometer with the contents of test tube 1, mix pairs of tubes (2 and 3, 4 and 5, 6 and 7, and 8 and 9) one pair at a time. After mixing one pair, measure the change in absorbance for two minutes at 20-second intervals for each temperature. The temperature will not remain exact, but the effects can be overlooked. Calculate the net change in absorbance from the 20-second mark to the 120-second mark. After measuring the absorbance changes, mix the second pair and measure the absorbance changes, and so on.

Note: The room-temperature experiment can be performed while the other tubes temperature-equilibrate.

Record changes in absorbance for the reaction mixture at each temperature in Table 2.2, and plot a histogram showing the change in absorbance at each temperature in Figure 2.4.

Table 2.1. Mixing table for temperature experiment (all values in milliliters)					
	Tube	Buffer (pH 5)	Catechol	Extract	Total Volume
	1	4.0	2.0	1	7
4°C	2	4.0	2.0	0	6
	3	0.0		1.0	1
23°C	4	4.0	2.0	0	6
	5	0.0		1.0	1
32°C	6	4.0	2.0	0	6
	7	0.0		1.0	1
48°C	8	4	2.0	0	6
	9	0.0		1.0	1

Table 2.2. Temperature effects on peroxidase activity (entries are absorbance units at 500 nm)				
Time (seconds)	Tubes 2 and 3 4°C	Tubes 4 and 5 23°C	Tubes 6 and 7 32°C	Tubes 8 and 9 48°C
Absorbance at start				
Absorbance 3 minutes later				
Change in absorbance				

pH Effects

Begin by stating null (H_0) and alternative (H_a) hypotheses that relate change in enzyme activity to the pH of the solution used:

H_0:

H_a:

To determine the effect of pH on peroxidase, perform the following experiment.

Your instructor will supply buffers at pHs of 3, 5, 7, and 9. Number test tubes 1 through 9. Setup pH-effect tests by adding the reagents described in table 2.3.

After adjusting the spectrophotometer with the contents of test tube 1, mix pairs of tubes one at a time (2 and 3, 4 and 5, 6 and 7, 8 and 9.) Measure absorbance changes at 20-second intervals for two minutes for each pair before mixing the next pair. Record the results in Table 2.4.

Table 2.3. Mixing table for pH experiment (all values in milliliters)					
pH	Tube	Buffer	Catechol	Extract	Total Volume
5	1	4.0 (pH 5)	2.0	1	7
3	2	4.0 (pH 3)	2.0	0	6
	3	0	0	1.0	1
5	4	4.0 (pH 5)	2.0	0	6
	5	0	0	1.0	1
7	6	4.0 (pH 7)	2.0	0	6
	7	0	0	1.0	1
9	8	4.0 (pH 9)	2.0	0	6
	9	0	0	1.0	1

Time (seconds)	Tubes 2 and 3 pH 3	Tubes 4 and 5 pH 5	Tubes 6 and 7 pH 7	Tubes 8 and 9 pH 9
Table 2.2. Effects of pH on peroxidase activity (entries are absorbance units at 500 nm)				
Absorbance at start				
Absorbance 3 minutes later				
Change in absorbance				

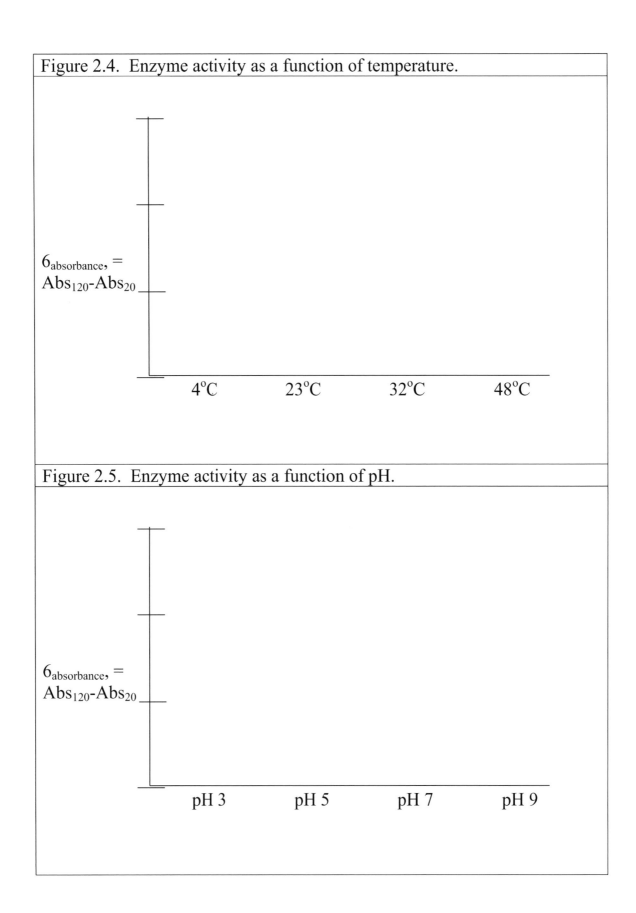

Figure 2.4. Enzyme activity as a function of temperature.

$6_{absorbance,} =$
$Abs_{120}-Abs_{20}$

4°C   23°C   32°C   48°C

Figure 2.5. Enzyme activity as a function of pH.

$6_{absorbance,} =$
$Abs_{120}-Abs_{20}$

pH 3   pH 5   pH 7   pH 9

References Cited:

Dolphin., Warren. 2002. *Biological Investigations: Form, Function, Diversity and Process*, edition 6. McGraw-Hill, Dubuque, Iowa.

Exercise 3: Use of Light Microscopes -- Compound and Stereoscopic (Dissecting)

(adapted from Lab Topic 2: Techniques in Microscopy, in Warren Dolphin's Biological Investigations, edition 6)

Supplies

Equipment
Compound microscope
Dissecting microscope

Materials
Prepared slides
      Compass
      Graph paper
Glass slides
Cover slips
toothpicks
0.8% saline
0.8% saline with methylene blue stain

Objectives
1) To learn the parts of a microscope and their functions
2) To investigate the optical properties of a light microscope, including image orientation, plane of focus, and measuring objects.
3) To understand the importance of magnification, resolution, and contrast in microscopy.

Background
Since an unaided eye cannot detect anything smaller than 0.1 mm (10^{-4} meters) in diameter, cells, tissues, and many small organisms are beyond our visual capability. A light microscope extends our vision a thousand times, so that objects as small as 0.2 micrometers (2×10^{-7} meters) in diameter can

be seen. (For comparison, red blood cells average 7 microns, i.e. micrometers, in diameter.) The electron microscope further extends our viewing capability down to 1 nanometer (10^{-9} meters). At this level, it is possible to see the outlines of individual protein or nucleic acid molecules.

Although 300 years have passed since its invention, the standard light microscope of today is based on the same principles of optics as microscopes of the past (Figure 3.1).

Figure 3.1. Anatomy of a typical (Olympus) binocular compound microscope.

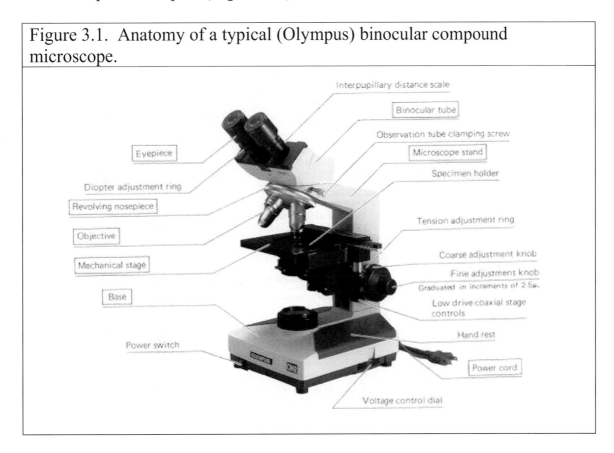

Microscope quality depends upon the capacity to **resolve**, not magnify, objects. The distinction between microscope resolution and magnification can best be illustrated by an analogy. If a photograph of a newspaper is taken from across a room, the photograph would be small, and it would be impossible to read the words. If the photograph were enlarged, or magnified, the image would be larger, but the print would *still* be unreadable, because it would now be too fuzzy. Regardless of the magnification used, the photograph would never make a fine enough distinction between the points on the printed page. Therefore, without

resolving power, or the ability to distinguish detail, magnification is worthless.

Modern microscopes increase both magnification and resolution matching the properties of the light source and precision lens components. Today's light microscopes are limited to practical magnifications in the range of 1000 to 2000x and to resolving powers of 0.2 micrometers. Most student microscopes have magnification powers to 450x or possibly to 980x, and resolving properties of about 0.5 micrometers. These limits are imposed by the expense of higher power objectives and the accurate alignment of the lens elements and light sources.

Avoiding hazards in microscopy

Use care in handling your microscope. The following list contains common problems, their causes, and how they can be avoided.

1) **Microscope dropped or ocular falls out**
   a. Carry microscope in upright position using BOTH hands, as shown in Figure 3.2);
   b. When placing the microscope on a table or in a cabinet, hold it close to the body; do not swing it at arm's length or set it down roughly;
   c. Position electric cords so that the micrsocope cannot be pulled off the table.
2) **Objective lens smashes coverslip and slide**
   a. Always examine a slide first with the low- or medium power objective;
   b. Never use the high-power objective to view thick specimens;
   c. Never focus downward with the coarse adjustment when using high-power objective.
3) **Image blurred**
   a. ALWAYS begin a viewing session by wiping all glass surfaces (ocular, objective, condenser, light) with lens paper.
   b. High-power objective was pushed through the coverslip 0see number 20 and lens is scratched;
   c. Slide was removed when high-powered objective was in place, scratching lens. Remove slide only when low-power objective is in place;

d. Use of paper towels, facial tissue, or handkerchiefs to clean objectives or oculars scratched the glass and ruined the lens. Use only *lens tissue* folded over at least twice to prevent skin oils from getting on the lens. Use distilled water to remove stubborn dirt;

e. Clean microscope lenses before and after use. Oils from eyelashes adhere to oculars, and wet-mount slides often encrust the objectives or substage condenser lens with salts;

4) Mechanical failure of focus mechanism
a. Never force an adjustment knob; this may strip gears;
b. Never try to take a microscope apart; you need a repair manual and proper tools.

Figure 3.2. Proper way to carry a microscope. USE BOTH HANDS. (Apparently, even TKE's can do it right.) From http://bio.winona.msus.edu/berg/IMAGES/scope9.jpg

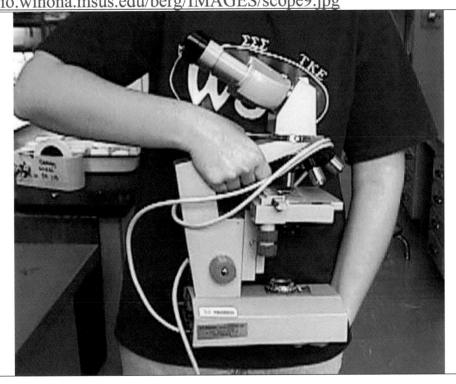

The Compound Microscope

⌘Get your microscope from its storage place, using the precautions just mentioned. Depending on its age, manufacturer, and cost, your compound microscope may have only some of the features discussed in this section. Look over your microscope and find the parts described, referring to Figure 3.1.

Ocular lens

The **ocular lens** is the lens you look through. If you microscope has one ocular, it is a **monocular** microscope. If it has two, it is **binocular**. Sometimes, there will be an additional attachment to accommodate a camera, so these will be **trinocular** microscopes. In binocular microscopes, one ocular is adjustable to compensate for the differences between your eyes. Ocular lens can be made with different magnifications. What magnification was stamped on your ocular lens housing?

Body tube

The **body tube** is the hollow housing through which light travels to the ocular. If the microscope has inclined oculars, the body tube contains a prism to bend the light rays around the corner.

Objective lenses

The **objective lenses** are a set of lenses mounted on a rotating **turret** at the bottom of the body tube. Rotate the turret and note the click as each objective comes into position. The objective gathers light from the specimen and projects it into the body tube. Magnification ability is stamped on each lens. They also may have a colored ring to distinguish each of the objectives. Complete the following table:

Size of objective	Magnification ability	Color of ring
Scanning (small)		
Low-power (medium)		
High-dry power (large)		
Oil immersion (largest)		

Stage

The horizontal surface on which the slide is placed is called the **stage**. It may be equipped with simple clips for holding the slide in place or with a **mechanical stage**, a geared device for precisely moving the slide. Two knobs, either on top of or under the stage, move the mechanical, either in a left-right (↔) or forward-backward (↕) direction.

Substage Condenser Lens

The substage **condenser** lens system, located immediately under the stage, focuses light on the specimen. An older microscope may have a mirror instead.

Diaphragm Control

The **diaphragm** is an adjustable light barrier built into the condenser. It may be either an **annular** or an **iris** type. With an annular control, a plate under the stage is rotated, placing open circles of different diameters in the light path to regulate the amount of light that passes to the specimen. With the iris control, a lever projecting from one side of the condenser opens and closes the diaphragm. **Which type of diaphragm does your microscope have?**

Use the smallest opening that does not interfere with the field of view. The condenser and diaphragm assembly may be adjusted vertically with a knob projecting to one side. Proper adjustment often yields a greatly improved view of the specimen.

Light source

The **light source** has an off/on switch and may have adjustable lamp intensities and color filters. To prolong lamp life, use medium to low (6 down to 3) voltages whenever possible. A second diaphragm may be found in the light source. If present, experiment with it to get the best image.

Base and Body Arm

The base and body arm are the heavy cast metal parts.

Coarse Focus Adjustment

Depending on the type of microscope, the **coarse adjustment knob** either raises or lowers the body tube or the stage quickly to focus the optics on the specimen. Use the coarse adjustment knob only with the scanning (4x) and low-power (10x) objectives. NEVER use it with the high-power (40x) objective.

Fine Adjustment

The **fine adjustment knob** changes to specimen-to-objective distance very slowly with each turn of the knob and is used for all focusing of the 40x and 100x objectives. It has no noticeable effect on the focus of the scanning objective (4x), and little effect when using the 10x objective.

The Compound Microscope Image

✠1) Magnification

Compound microscopes consist of two lens systems; the objective lens, which magnifies and projects a "virtual image" into the body tube, and the ocular lens, which magnifies that image further and projects the enlarged image into the eye.

The total magnification with each objective can be calculated by multiplying the magnifying power of the ocular by the magnifying power of the objective. **What magnifications are possible with your microscope?**

Objective	Magnifying ability of the objective	Magnifying power of the ocular	Total magnification (equals product of magnifying abilities)
Scanning power			
Low power			
High power			
Oil immersion			

✖ 2) Brightness

Plug in your microscope and click the scanning power objective into place. Look through the ocular and you will see a circular window of light. Now click the oil immersion objective into place. **Which objective allows more light through?**

✖ 3. Image orientation

With the scanning objective in place, observe the "Compass" slide through the compound microscope and then with the naked eye. Draw what you see in Figure 3.5a. Is there a difference in the orientation of the letters between the object and the image? While looking through the microscope, try to move the slide so that the image moves to the left. Which way did you have to move the slide? Try to move the image down. Which way did you have to move the slide.

Now place the slide on the dissecting microscope. Draw what you see in Figure 3.5b. Is there a difference in the orientation of the letters between the object and the image?

Figure 3.3. Comparison of object and images projected by compound and dissecting microscopes.	
The object: A piece of newsprint with the compass points labeled:	
N **W O E** **S**	
Figure 3.5a. Orientation of the letters in the newsprint through the compound microscope.	Figure 3.5b. Orientation of the letters in the newsprint through the dissecting microscope.

4) Estimating the size of the field of view

Although using a standardized **ocular micrometer** will allow exact measurement of microscope structures, it is possible to estimate the size of the field of view by using a slide of millimeter-square graph paper.

First, click the scanning (4x) objective into place, and then place a slide of the graph paper on the mechanical stage. Use the knobs of the mechanical stage to position the graph paper in such a way that there is a horizontal line across the middle of the field-of-view, and there is a vertical line at the extreme left-hand side of the field, as in the following figure:

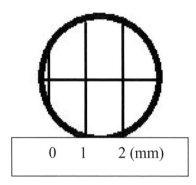

You can use the visible section of the graph paper like a ruler, where the vertical line at the extreme left is your "0" mark, and the subsequent vertical lines mark off millimeters. Estimate the total size of the field of view when the total magnification is 40x.

Now click the 10x objective into place, and use the knobs of the mechanical stage to position the graph paper in such a way that there is a horizontal line across the middle of the field-of-view, and there is a vertical line at the extreme left-hand side of the field, like you did before. You will notice that the size of the field-of-view is smaller, and the width of the lines is wider. That is because you are now looking at a smaller window. Estimate the total size of the field of view when the total magnification is 100x.

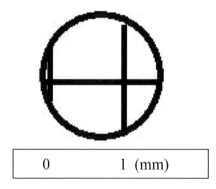

Finally, click the 40x objective into place, and use the knobs of the mechanical stage to position the graph paper in such a way that there is a horizontal line across the middle of the field-of-view, and there is a vertical line at the extreme left-hand side of the field, as you did twice before. You will notice that there is a very uneven distribution of ink marking the lines, and you will notice that there is no subsequent vertical line. That means that at 400x, the field of view is less than 1 mm in diameter, as shown in the following figure:

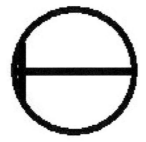

The size of the field of view can be estimated by using the following formula:

$$\frac{100 \times (\text{Diameter at 100x})}{400} = \text{Diameter at 400x}$$

Fill in the following table, which will provide you with an estimate of the field-of-view for each of the levels of magnification that we will be using in this class:

Magnification	Diameter of the Field-of-view (mm)
40x	
100x	
40x	

Stereoscopic Dissecting Microscopes

The stereoscopic microscope, usually called a **dissecting microscope**, differs from the compound microscope in that it has two (rather than one) objective lenses for each magnification. This type of microscope always has two oculars, and it is used to observe objects in 3 dimensions. The resolution and magnification capabilities are less than in a compound microscope, and the magnifications on this type of microscope usually range from 4x to 50x (Figure 3.4). You've already seen that the image projected by a dissecting microscope is different than the one projected by compound microscopes.

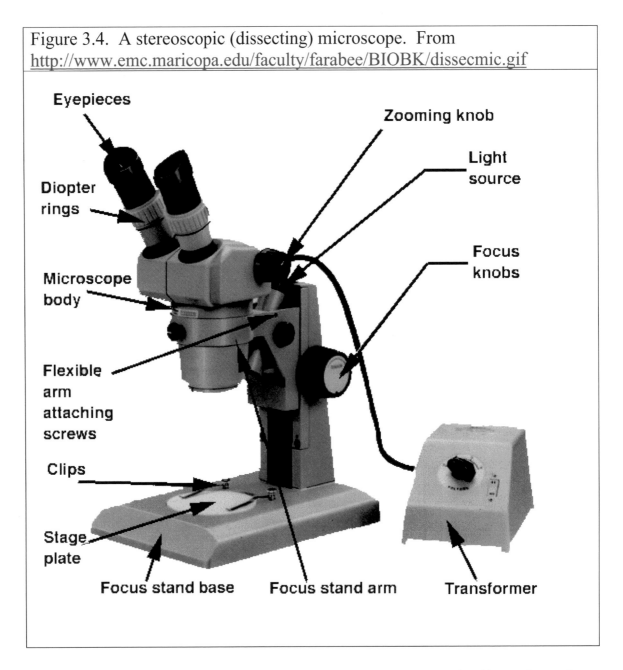

Figure 3.4. A stereoscopic (dissecting) microscope. From http://www.emc.maricopa.edu/faculty/farabee/BIOBK/dissecmic.gif

Stereoscopic microscopes are often used for the microscopic dissection of specimens. The light source may come from above the specimen and be reflected back into the microscope, or it may come from underneath and be transmitted through the specimen into the objectives. The stage may be clear glass or an opaque plate, white on one side and black on the other. The choice of illumination depends on the task to be performed and on whether the specimen is opaque or translucent.

✠ **Observations of spinal cord cross section images with the compound and dissecting microscopes.**

Obtain a slide showing the cross-section of a spinal cord, and mount it on the dissecting microscope. Bear in mind that the total magnifying power of the dissecting microscope is limited -- it will be between 4x and 40x. Even at 40x magnification, it will not be possible to view individual neurons in the gray matter (Figure 3.5).

Figure 3.5. View of a spinal cord cross-section through a dissecting microscope.	
a. Grey matter appears a lighter color, while white matter appears charcoal. The white areas are an artifact of the slide-making process.	b. Draw the spinal cord as you see it through a dissecting microscope. Label the white matter and gray matter.

Now, mount the slide onto the compound microscope, and look at the gray matter under 100x. At this level of magnification, you may be able to see individual neurons. Draw and label the various structures in the spinal cord (Figure 3.6).

Figure 3.6. View of a spinal cord under low (40x or 100x) power.	
a. View of the spinal cord, showing junction between white matter (stained charcoal) and grey matter (tan-like color)	b. Draw the spinal cord as you see it through a compound microscope under 40x or 100x. Label the white matter and gray matter.
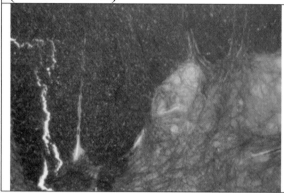	

Now, click on the 40x objective into place. AT THIS MAGNIFICATION, USE THE FINE ADJUSTMENT KNOB. At this level of magnification, you should be able to see the nuclei and nucleoli of neurons, as well as the axons emerging from the cell bodies (Figure 3.7).

<table>
<tr><td colspan="2">Figure 3.7. View of the spinal cord under 400x with the compound microscope.</td></tr>
<tr><td>a. View of motor neurons from a spinal cord.</td><td>b. Draw the spinal cord or individual neurons as you see it through a compound microscope under 400x. If possible, label cell structures.</td></tr>
<tr><td>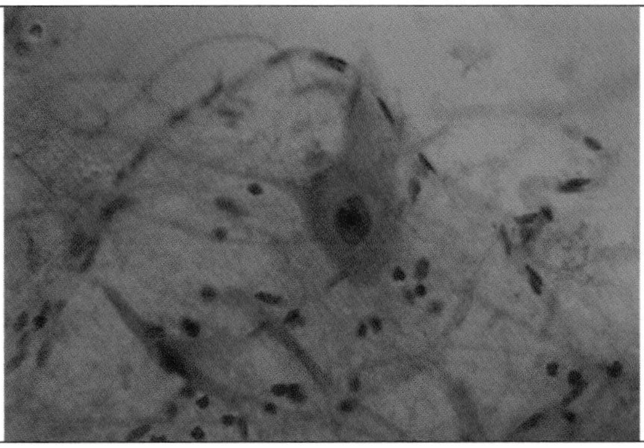</td><td></td></tr>
</table>

Preparation of a wet mount to show simple, squamous epithelial tissue from the inside surface of the buccal cavity.

Methylene blue dye is frequently used to highlight the nucleus in wet mounts of cells extracted from the buccal cavity, i.e. the inside surface of the mouth. Take a clean microscope slide, and place two drops of methylene blue stain the middle. Then get a wooden toothpick, and **gently** scrape the inside surface of your mouth, which is lined with a type of epithelial tissue called **simple, squamous epithelial** tissue.

Although you may not see anything at the tip of the toothpick, these cells are easily dislodged. In fact, the lining of the entire digestive system is replaced every 24 hours. Now roll the tip of the toothpick in the methylene blue drops. Discard the toothpick, and then look at YOUR VERY OWN cheek cells under the microscope. First scan the slide at 40x, and then go up to 100x. At this level of magnification, you should be able to see clumps of cells floating in the dye. Finally, by going up to 400x magnification, you should be able to see individual cells and their nuclei (Figure 3.8).

Figure 3.8. Wet mount of cells extracted from the buccal cavity, and stained with methylene blue.	
a. What they should look like, under 400x.	b. Draw a representative cell (or 2) as you see them through a compound microscope under 400x. If possible, label cell structures.
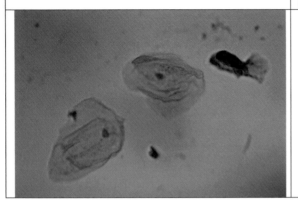	

Exercise 4: Fetal Pig Dissection

(Adapted from Warren D. Dolphin's Biological Investigation: Form, Function, Diversity, and Process, edition 6)

Supplies

Equipment

Dissecting microscopes
Dissecting pans
Dissecting tools

Supplies

Double-injected fetal pigs
Plastic bags for storage

General Dissection Information

Fetal pigs are unborn fetuses taken from a sow's uterus when she is slaughtered. They are a by-product of meat preparation and are used in teaching basic mammalian anatomy. Often the circulatory system has been injected with latex so that the veins will appear blue and the arteries red. The blue latex rubber is administered via the superior vena cava, so that explains the slash in the neck of the fetal pig. To highlight the arterial system, red latex rubber is injected into the umbilical arteries.

Strong preservatives are used and can irritate your skin, eyes, and nose. Rinse your pig with tap water to remove some of the preservative to lessen irritation. If you wear contact lenses, you may want to remove them during dissection, since the preservative vapor can collect in the water behind the lens and can be very irritating. If a lanolin-based hand cream is available, use it on your hands before and after dissection to prevent drying and cracking of the skin. Alternatively, your instructor may ask you to purchase rubber gloves and use them during this and subsequent dissections.

You will use this same fetal pig in several future labs to study the circulatory, respiratory, and excretory systems. This means that you must do a careful dissection each time so that as many structures as possible are left undamaged and in their natural positions. Two good rules to keep in mind as you dissect your animal are: *cut as little as possible and never remove an organ unless you are told to do so.* If you indiscriminately cut into the pig to find a single structure without regard to other organs, you will undoubtedly ruin your animal for use in future laboratories.

Many of the instructions for dissection in this lab and later ones use anatomical terms to indicate direction and spatial relationships when the animal is alive in normal orientation. You should know the meaning of such terms as:

Anterior - situated near head or, in animals without heads, the end that moves forward.

Caudal - extending toward or located near tail.

Cephalic - extending toward or located on or near head (also cranial).

Distal - located away from the center of the body, the origin, or the point of attachment.

Dorsal - pertaining to the back as opposed to **ventral**, which pertains to the belly or lower surface.

Median - a plane passing through a bilaterally symmetrical animal that divides it into right and left halves.

Posterior - toward the animal's hind end: opposite of anterior.

Proximal - opposite of distal

Right-left - always in relation to the animal's right and left, not yours.

Sagittal - planes diving an animal along the median line or parallel to the median.

Obtain a fetal pig and place it in a dissecting pan or tray, ventral side up. Take two pieces of string and tie them tightly to the ankles of both right legs (the animal's right legs, not the legs to your right when the pig is lying on its back.) Run the strings under the pan and tie each to the corresponding left leg. Stretch the legs to spread in internal organs for easier dissection (Figure 4.1). Do not pull so hard that you break internal blood vessels.

External anatomy of fetal pig

Rows of **mammary glands** and the **umbilical cord** should be prominent on the ventral surface of your pig. Mammary glands and hair are two of the diagnostic characteristics of the class *Mammalia*. The umbilical cord attached the fetal pig to the placenta on the sow's uterus. Look at the cut end ef the cord and note the blood vessels that carried nutrients, wastes, and dissolved gases from the fetus to the placenta, where they were exchanged by diffusion with the maternal circulatory system.

Determine the sex of your pig. Identify the **anus**. In females, there will be a second opening, the **urogenital opening**, ventral to the anus. In males, the urogenital opening is located just posterior to the umbilical cord. **Scrotal sacs** will be visible just ventral to the anus in males.

Figure 4.1 shows the sequence of cuts that should be made to expose the internal organs. Make the cuts in the sequence indicated.

Mammalian digestive system

Anatomy of the mouth

With heavy scissors, a razor blade, or scalpel, cut through the corners of the mouth and extend the cut to a point below and caudal to the eye.

Open the mouth, as in Figure 4.2, and observe the **hard palate**, composed of bond covered with mucous membrane, and the **soft palate**, which is a caudal continuation of the soft tissue covering the hard palate. The oral cavity ends and the **pharynx** starts at the base of the tongue.

The pharynx is a common passageway for the digestive and respiratory tracts. The opening to the **esophagus** may be found by passing a blunt probe down along the back of the pharynx on the midline. This collapsible tube connects the pharynx with the stomach. The **glottis** is the opening into the **trachea** or windpipe and lies ventral to the esophagus. It is covered by a small white tab of cartilage, the **epiglottis**. The epiglottis may be hidden from view in the throat; if so, you will have to pull it forward with forceps or a probe to see it.

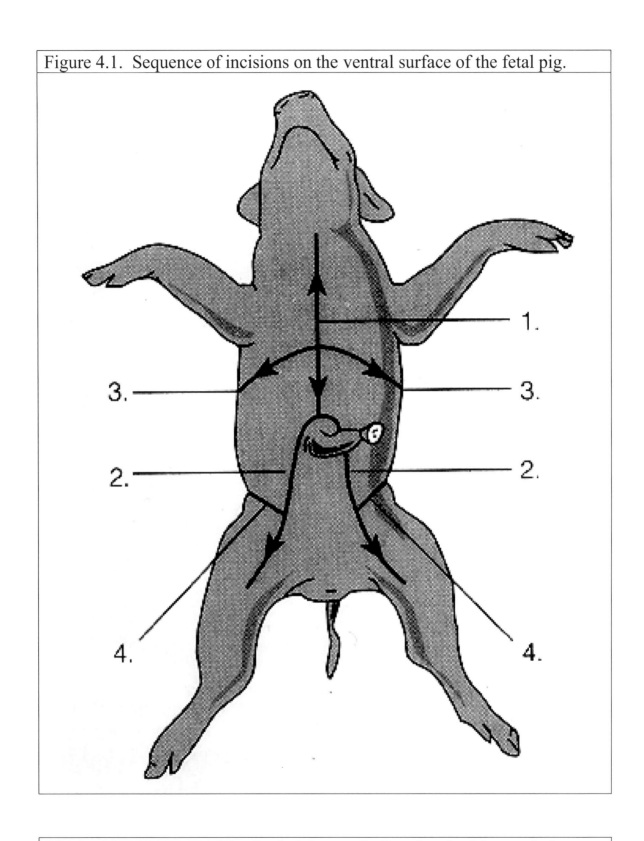

Figure 4.1. Sequence of incisions on the ventral surface of the fetal pig.

1.

3.   3.

2.   2.

4.   4.

Figure 4.2. Anatomy of the fetal pig's mouth. Reprinted with permission from Dr. Mary C. Flath from

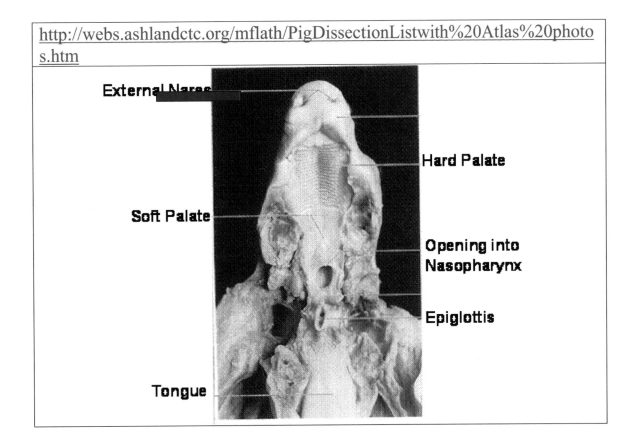

Alimentary Canal Anatomy

We can begin our dissection with a virtual tour of the visceral organs found with the fetal pig via
http://bioweb.wku.edu/courses/biol225/images/Zoolab5.htm

You will see the path of the esophagus to the stomach when you dissect the respiratory system. To view the rest of the alimentary tract and associated glands, use a scalpel or pair of scissors to make incisions into the **abdominal** cavity, as indicated in Figure 4.1. Cut carefully through only the skin and muscles to avoid damaging the internal organs.

As you lay the tissue flaps back, you will see the organs of the abdominal cavity covered by a translucent **peritoneal** membrane. The flap containing the umbilical cord will be held in place by blood vessels. Tie both ends of a 15-com piece of thread to the blood vessel about 1 cm apart. Cut the vessels between the two knots and lay this tissue flap back. Leave the thread in place so you later can trace the circulatory system.

Find the thin, transparent membranes, the **mesenteries**, which suspend and support the internal organs (**viscera**) in the body cavity. The dark brown, multilobed **liver** should be visible caudal to the **diaphragm** (Figure 4.3). If you trace the umbilical vein from the thread to the liver, you will see a green-colored sac, the **gallbladder**, located just below the entrance of the vein into the liver. It stores bile produced in the liver. Bile travels from the gallbladder to the small intestine via the bile duct. Bile is an emulsifying agent that aids in digestion of fats.

Under the liver on the left side is the **stomach**. Locate the point where the esophagus enters the **cardiac region** of the stomach. Gastric glands in the wall of the stomach secrete pepsinogen, hydrochloric acid, and rennin. Pelsinogen is activated by hydrochloric acid to become pepsin, which digests proteins. Rennin is an enzyme that hydrolyzes milk protein. Food leaves the stomach as a fluid suspension, chyme. It enters the **duodenum**, the first part of the small intestine.

Find the **pancreas**, a glandular mass lying in the angle between the curve of the stomach and duodenum. It secretes several enzymes into the duodenum that digest proteins, lipid, carbohydrates, and nucleic acids. Certain cells in the pancreas act as endocrine cells and secrete the hormones insulin and glucagon. In fact, insulin used in human diabetes therapy can be extracted from the pancreases of pigs collected at slaughterhouses.

Remove the stomach by cutting the esophagus and duodenum. Slit the stomach lengthwise, cutting through the cardiac and pyloric **sphincters**, muscles that regulate passage of material into and out of the stomach. The internal surface of the stomach is covered by gastric mucosal cells, which secrete mucus that prevents the stomach from digesting itself. When this protection fails, a peptic ulcer develops.

The small intestine is made up of three sequentially arranged regions: **duodenum**, **jejenum**, and **ileum**. These areas are difficult to differentiate from each other. Cut out a 2-cm section of the small intestine about 10 cm posterior from the stomach, slit it open, and place it under water in a dish. Use your dissecting microscope to observe the velvety internal lining made up of numerous fingerlike projections called **villi**. The villi are highly vascularized, containing capillaries and lymphatics that transport the products of digestion to other parts of the body, especially the liver.

The ileum opens into the large intestine, or **colon**. They join at an angle, forming a blind pouch, the **cecum**, which in primates and some other mammals often ends in a slender appendage, the **appendix**. In herbivores, the cecum is very large and contains microorganisms that aid digestion by breaking down cellulose.

The **rectum** is the caudal part of the large intestine, where compacted, undigested food material is temporarily stored before being released through the **anus**. The colon of vertebrates contains large number of symbiotic bacteria, epecially *Escherichia coli*.

Figure 4.3. Major organs in the fetal pig (note that the stomach is not visible. It can be seen by raising the lower edge of the liver.) Reprinted with permission from Dr. Mary C. Flath from http://webs.ashlandctc.org/mflath/PigDissectionListwith%20Atlas%20photos.htm

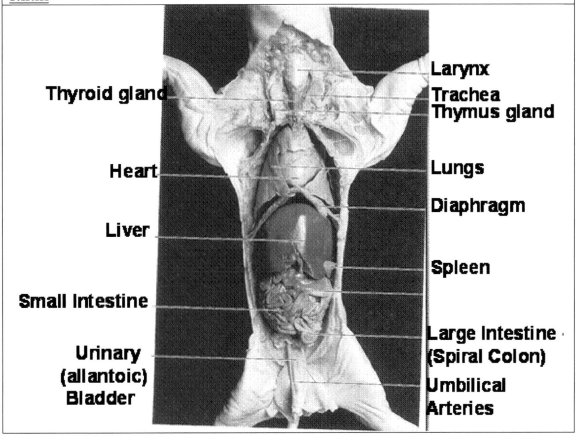

Histology of small intestine

Put your fetal pig aside and obtain a prepared slide of a cross-section of a mammalian small intestine. Examine it under scanning power (40x) and higher power (100x) with the compound microscope and compare with Figures 4.4.

Figure 4.4. Photomicrograph of a cross-section of mammalian small intestine, showing structure of villi.

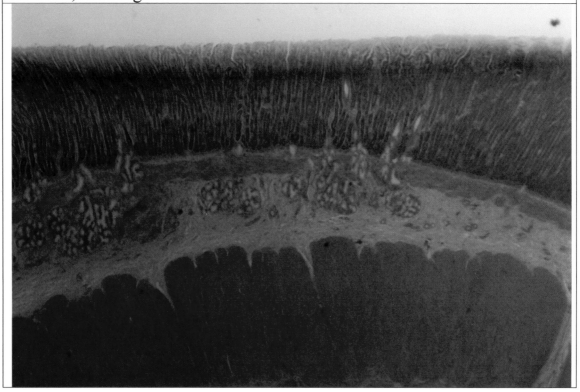

Figure 4.5. Close-up of villi.

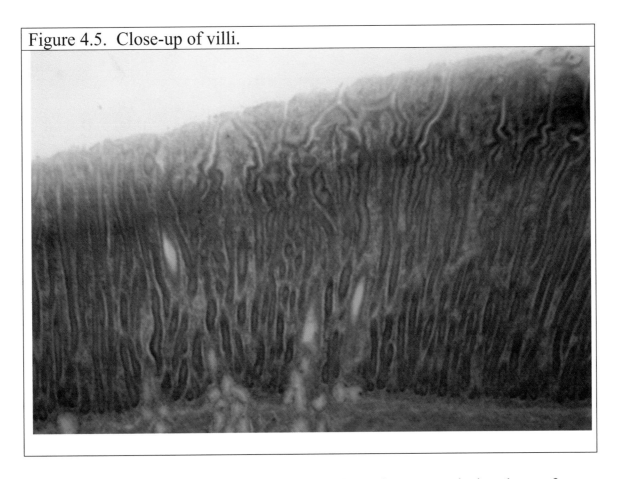

Villi increase the surface area of the small intestine to maximize the surface-area-to volume ratio (Figures 4.6). Bear in mind that the mode of entry of most drugs is via the digestive system.

Figure 4.6. Light photomicrograph of an *Amphiuma* small intestine, cut in cross-section.

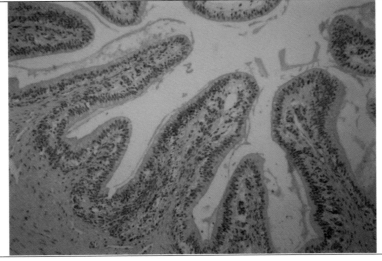

The central opening of the small intestine is called the **lumen** and is the space through which food passes as chyme during digestion. You can observe the small fingerlike projections (villi) of the intestine's inner surface. These villi are covered by a layer of cells called the **mucosa**. You should be able to distinguish two cell types in the intestinal mucosa: **goblet cells** and **columnar epithelial cells**. Examine them with the high-power objective. The goblet cells secrete mucus into the small intestine, serving as a lubricant for the passage of chyme. Epithelial cells are involved in absorption.

Return to the low-power objective and observe the **submucosa**, a layer of connective tissue that underlies the mucosa. Look for the blood vessels and lymphatic vessels that ramify through this layer. Sugars, amino acids, glycerides, and other components of digested food must move through the mucosal cells into the submucosa before they can enter the circulatory system and be distributed throughout the body.

To the outside of the submucosa are two smooth muscle layers; an **inner circular layer** and **outer longitudinal layer**. The inner circular muscles change the diameter of the intestine, and the outer muscles alter its length. These muscles contract in a wavelike motion called **peristalsis**, which pushes chyme through the digestive tract. The small intestine is covered by a layer of peritoneal cells, that together with underlying connective tissue, is called the **serosa**.

Exercise 5: Nervous system: Structure and function of the brain and spinal cord

(Adapted from Warren Dolphin's Biological Investigations: Form, Function, Diversity and Process, edition 6.)

Supplies

Compound microscope
Dissecting microscope
Oscilloscope
Nerve chamber
Stimulator

Materials

Prepared slides
      Spinal cord smear
      Spinal cord cross-section
Sheep brains
      Whole
      Sagitally-sectioned half-brains
Human brains
      Sagitally & coronally-sectioned preserved brains
      Models
Dissecting pans, tools
1 liter of 6% ethanol, for anesthetizing earthworms
Live earthworms

Background

Coordination of the several different types of specialized tissues in multicellular animals is necessary for the organism to operate as an integrated whole. The nervous system coordinates the body's relatively rapid responses to changes in the environment. The endocrine system regulates longer term adaptive responses to changes in body chemistry between meals, as the seasons change, or as developmental changes occur during maturation. Because the functions of these two systems often complement one another, biologists often speak of the **neuroendocrine system**.

The nervous system has three functions: (1) to receive signals from the environment and from within the body through the sense organs; (2) to process the information received, which can involve integration, modulation, learning, and memory, and (3) to cause a response in appropriate muscles or glands. **Receptors** are usually specialized cells outside of the central nervous system that detect physical and chemical changes. There are separate receptor cells for heat, cold, light, and so on. The function of the receptor cells is to convert the environmental signal into a change in the cellular membrane's ion permeability, leading to a voltage change across the cell membrane.

The resting potential of a neuron is -70 mv (millivolts), indicating that the cytoplasm is negatively charged with respect to the interstitial fluid bathing the neurons. The voltage across the cell membrane is due to a gradient of ions across the membrane, as indicated by Table 5.1:

Table 5.1. Comparison of ion concentrations, in mM				
	Na+ (sodium)	**K+ (potassium)**	**Cl- (chloride)**	**A- (anions, including proteins and amino acids)**
Inside neuron	10	150	4	100
Outside neuron	142	5	103	~0

The function of the receptor cells is to convert the environmental signal into a change in the cellular membrane's ion permeability, leading to a voltage change across the cell membrane. If the voltage change is of sufficient magnitude, an **action potential** will be created in an adjacent neuron, and a nerve impulse will travel from the sensing zone to the central nervous system

Histology and gross anatomy of the spinal cord

✠ Obtain a spinal cord smear. With this type of slide, the tissues have been made brittle so that with light pressure, the cells making up the spinal cord are separated. The integrity of the spinal cord is thereby lost, but we will be able to see the distinct sections of the spinal in a spinal cord cross-section slide.

You should be able to see both relatively large interneurons, with prominent nuclei, along with relatively minute glial cells, as in Figure 5.3:

| Figure 5.3. Spinal cord smear preparation. |
| Photomicrograph of a spinal cord smear. Note the relative size of the motor neuron and surrounding neuroglial cells. |

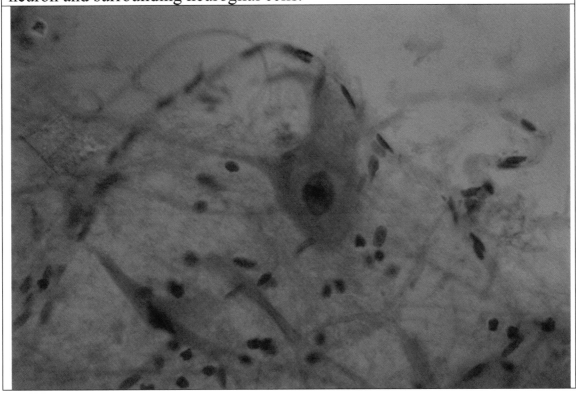

◈Now obtain a slide of a spinal cord cross-section. In this preparation, the gray matter, where the motor neuron and interneuron cell bodies are located, and the surrounding white matter, where the motor neuron axons are located, stain and appear different (Figure 5.4).

◈The spinal cord of the fetal pig can be exposed by laying the pig on its stomach and exposing the **spine** by removing the muscles that cover and extend on each side of the vertebrae. Since it is time consuming for each student to expose the entire spine, groups in the laboratory should expose only short sections of about 5 to 8 vertebrae each. The appearance of the spinal cord cross-section will be different at different locations in the spinal cord.

Figure 5.4. Spinal cord cross-section

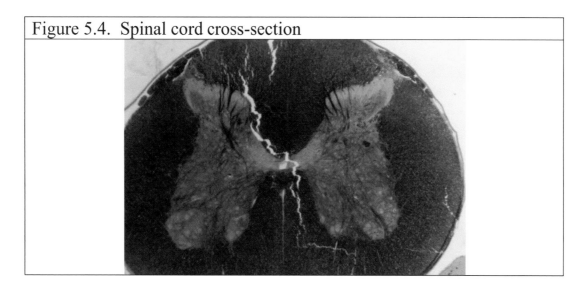

Synaptic transmission

Transmission between neurons or between a neuron to an effector cell is mediated through synapses, spaces through which neurotransmitter molecules are released from the presynaptic cell from vesicles in the presynaptic terminal (Figure 5.5) to receptors on the membrane of the postsynaptic cell. Binding of the neurotransmitter molecules to receptor molecules on the membrane of the postsynaptic cell will induce internal changes.

Figure 5.5. Model of a presynaptic terminal. The spherical objects that resemble bits of KIX cereal are vesicles containing neurotransmitter molecules. This process requires a considerable amount of energy, as reflected by the prominent mitochondrion.

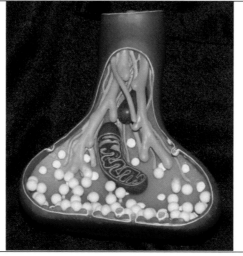

Gross anatomy of the brain

Functions of the brain are localized, which means that specific parts of the brain control specific functions (Table 5.2).

Table 5.2. Functional areas of the brain (adapted from Solomon, Berg and Martin, 1999)

Structure	Function
Medulla	Contains vital centers (clusters of neuron cell bodies) that control pulse, respiration, and blood pressure; contains centers that control swallowing, coughing, vomiting.
Pons	Connects various parts of brain with one another, contains respiratory center.
Midbrain	Center for visual and auditory reflexes (e.g. pupil reflex, blinking, adjusting ear to volume of sound)
Thalamus	Main sensory relay center for conducting information between spinal cored and cerebrum. Neurons in thalamus sort and interpret all incoming sensory information (except olfaction) before relaying messages to appropriate neurons in cerebrum.
Hypothalamus	Contains centers for control of body temperature, appetite, fat metabolism, and certain emotions; regulates pituitary gland.
Cerebellum	Reflex center for muscular coordination and refinement of movements; when it is injured, performance of voluntary movements is uncoordinated and clumsy.
Cerebrum	Motor functions, Receives information from receptors in the skin, Interprets visual input, Interprets auditory and olfactory information, Speech, Recognition and interpretation of words

Use the sheep whole- and half- brains, the sections of preserved human brains, and the human brain models to visualize the structures listed in Figures 5.6a through 5.6d.

Figure 5.6. Sheep and human brain images. Both sheep brain images are from http://darwin.baruch.cuny.edu/gelfond/Bio1005/neural.htm

a. Labeled sheep half-brain

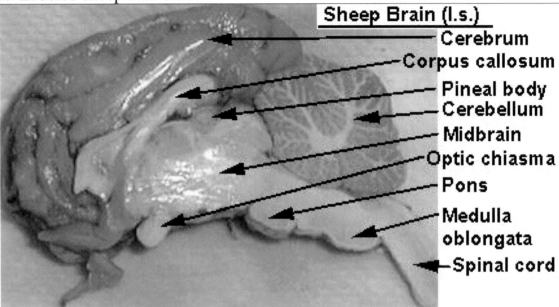

b. Labeled sheep whole brain

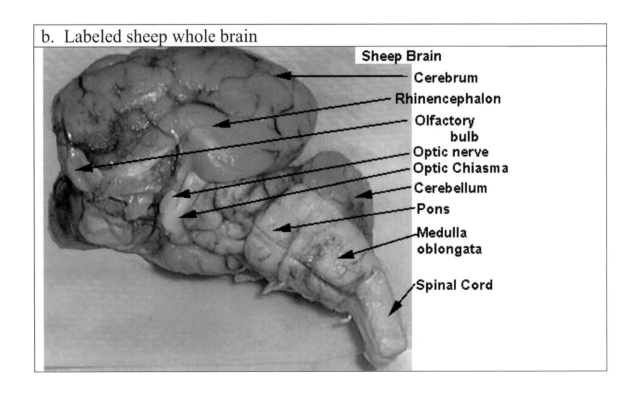

c. Normal human brain exterior view

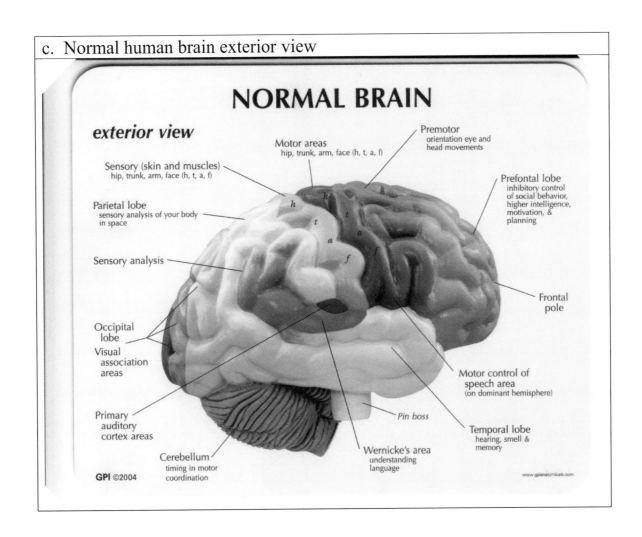

NORMAL BRAIN

exterior view

Sensory (skin and muscles)
hip, trunk, arm, face (h, t, a, f)

Motor areas
hip, trunk, arm, face (h, t, a, f)

Premotor
orientation eye and
head movements

Parietal lobe
sensory analysis of your body
in space

Prefontal lobe
inhibitory control
of social behavior,
higher intelligence,
motivation, &
planning

Sensory analysis

Frontal
pole

Occipital
lobe

Visual
association
areas

Motor control of
speech area
(on dominant hemisphere)

Primary
auditory
cortex areas

Pin boss

Temporal lobe
hearing, smell &
memory

Cerebellum
timing in motor
coordination

Wernicke's area
understanding
language

GPI ©2004

www.gpianatomicals.com

59

d. Normal human brain: interior view

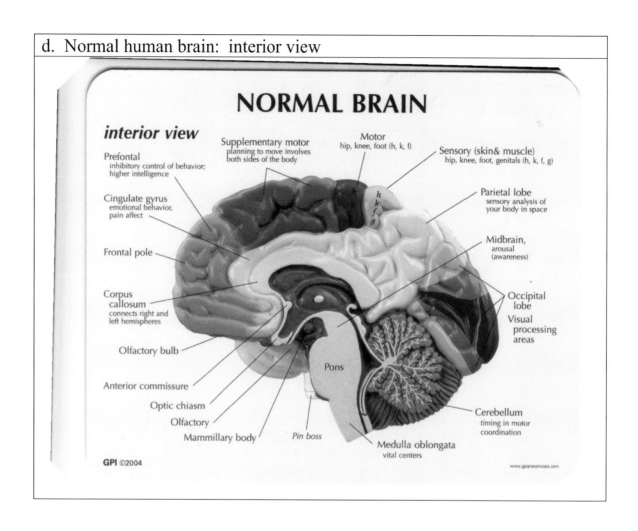

Exercise 6: Respiratory System, and assessment of the effects of cigarette smoking and athleticism on vital capacity.

(adapted from Warren Dolphin's Biological Investigations, edition 6)

Equipment
Compound microscope
Dissecting microscope
Spirometer
Computer with Excel software for analysis of vital capacity data
Dissecting tools

Supplies
Fetal pigs
Lungs-preserved and/or freeze-dried
Lungs-models
Prepared slides
       Lung cross-section
       Trachea cross-section
Vital capacity database of previous classes

Background
Cells of most animals are capable of aerobic metabolism. Animals, therefore, must have some mechanism for exchanging gases with its environment. Oxygen must move from the environment to every cell in the body, where it ultimately functions as a terminal electron (hydrogen) acceptor. Without oxygen, the mitochondrial cytochrome system would not operate, and cells could not carry out glycolysis, the Krebs cycle, and other oxidations of food materials. Furthermore, carbon dioxide, which is produced during the Krebs cycle and in other reactions, must pass from cells to the environment to maintain a consistent intracellular acid-base balance.

In small organisms with relatively low metabolic rates and large surface areas relative to body mass, free diffusion of these gases satisfies an animal's needs. Protozoa, sponges, cnidarians, flatworms, and roundworms have no anatomical specializations for respiration. Gases simply pass to and from the environment through their surface layers.

Larger aquatic animals, such as lobsters, clams, and fish, have developed specialized respiratory surfaces called gills. These featherlike surfaces allow the body fluids to circulate in a closed network that is separated by only a cell layer from the water in the environment. As the blood flows through the gills in one direction, water passes over the external gill surface in the opposite direction, allowing a very efficient exchange of respiratory gases.

The oxygen-carrying capacity of blood is increased by the presence of a respiratory pigment, such as **hemocyanin** (usually dissolved in hemolymph) or **hemoglobin** (usually in blood cells). These pigments bind loosely and reversibly with oxygen to facilitate oxygen transport from the gills to the tissues. Blood is oxygen deficient when it enters the gills or lungs. Oxygen diffuses into the aqueous portion of the blood and combines with the pigments. As the blood flows to areas where there is little oxygen, the diffusion gradient causes the oxygen to move into the tissue, where it is used by mitochondria. Conversely, a high concentration of carbon dioxide is usually present in active tissues. The CO_2 moves down the diffusion gradient into the blood where it simply may dissolve or may combine chemically with the pigments. The carbon dioxide is, in turn, carried back to the respiratory organs where it is exchanged with the environment.

Gross anatomy of the respiratory system

The human respiratory system is shown in Figure 6.1. The respiratory system is often referred to as a respiratory tree. Note that the respiratory tree is entirely hollow, and is inverted. The models of the human respiratory system are life-sized.

We will be continuing the dissection of the fetal pig to examine its respiratory system in greater detail.

Figure 6.1. Model of the human respiratory system, showing trachea, bronchi, and lobes of lungs.

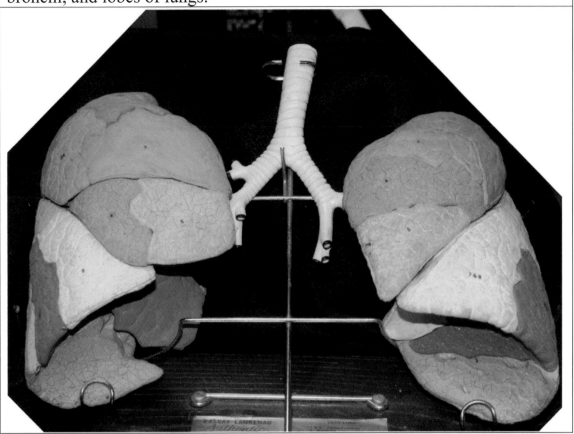

✠Return to your fetal pig and examine the external openings (**external nares**) on the snout. Cut across the snout with a scalpel about 1 to 2 cm from the end and remove the tip. The nasal passages are separated from each other by the **nasal septum**. The curved **turbinate bones** in the sinus area incrase the surface area of the passageways, creating eddy currents that, along with hairs, cilia, and mucus, help remove dust in the inhaled air and humidify it. The floor of the nasal passages is made up of the hard palate and the soft palate (posterior to hard palate).

✠Look into the pig's mouth. Behind and above the soft palate is the **nasopharynx**. It may be necessary to slit the soft palate to observe this. Air enters the nasopharynx from the posterior end of the nasal passages, then passes into the pharynx, through the glottis, and into the larynx and ultimately the trachea. If food accidentally enters the glottis, choking results.

✠In the nasopharyngeal area, look for the openings of the **eustachian tubes**. They are very difficult to find. These tubes allow air pressure to equilibrate between the middle ear chamber and the atmosphere. (This is why changes in altitude cause the ears to "pop".) Throat infections often spread in the ears through the eustachian tubes.

✠Run your fingers over the pig's throat and locate the hard, round **larynx**. Make a **medial** incision in the skin of the throat and extend the end cuts laterally, folding back the skin flaps. Repeat this procedure for the muscle layers.

✠Note the large mass of glandular material, the **thymus**, in this area. In the young pig, the thymus produces **lymphocytes**, an important component of the immune system. Later in life, the thymus atrophies and is of little consequence.

As you approach the larynx and trachea in your dissection, use a blunt probe to separate the muscles and expose these structures. Ventral to the trachea, observe the brownish-colored **thyroid gland**. Note how both the trachea and larynx are supported by rings of cartilage. The **hyoid apparatus** is anterior to the larynx and is a supporting frame for the tongue extensor muscles.

The larynx, or voice box, contains folks of elastic tissue, which are stretched across the cavity. These **vocal cords** vibrate when air passes over them, producing sound, and attached muscles vary the cord tension, allowing variations in pitch. Slit the larynx longitudinally and observe the vocal cords. Continue the slit posteriorly into the trachea and observe its lining. The esophagus is located behind the trachea. Pass a blunt probe into the esophagus from the mouth to help identify it.

✠If the **thorax** of your animal is not already opened, make a longitudinal cut with heavy scissors through the ribs just to the right of the **sternum**, or breastbone. Always keep the lower scissor tip pointed upward against the inside of the sternum to avoid catching and cutting internal structures.

The **diaphragm** is a sheet of muscle that separates the **abdominal cavity** from the thoracic cavity. The thoracic cavity is divided into three areas by membranes: the right and left **pleural cavities**, which surround the lungs, and the **pericardial cavity** where the heart is located.

If the pleural membranes are removed, the **lung** structure can be seen. The trachea, when it enters the thorax, divides into two **bronchi**, which are hidden from direct view beneath the heart and blood vessels. These bronchi, which finally end in microscopic air sacs called **alveoli**. Alveoli have walls only a single cell layer thick and they are covered by capillaries. In these air sacs, oxygen and carbon dioxide are exchanged between the blood and the inhaled air.

Histology of the respiratory system

✠ Remove a piece of the lung and put it in a small bowl of water. Observe it with a dissecting microscope and find the alveoli and bronchioles.

✠ Now look at Figure 6.2 and a prepared slide showing a cross-section of alveolar tissue. How many thicknesses separate the air in a mammalian lung from the red blood cells in the capillaries?

Figure 6.2. Cross-section of alveoli

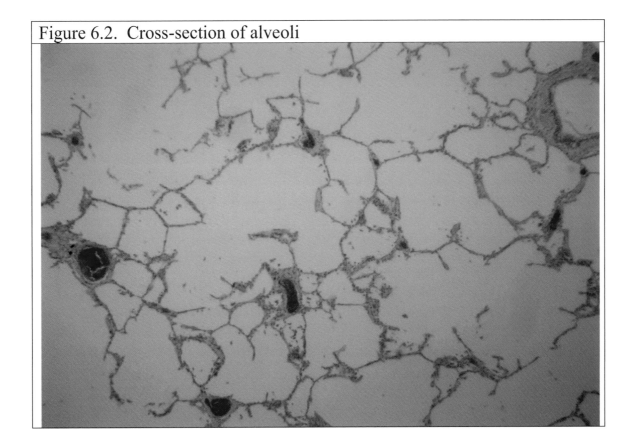

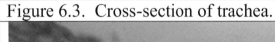Now obtain a prepared showing the cross-section of a trachea, as in Figure 6.3. Note that the entire structure consists of several distinct layers. Air is in contact with the pseudostratified ciliated columnar epithelial tissue is it travels from the nasopharynx to the lungs.

Figure 6.3. Cross-section of trachea.

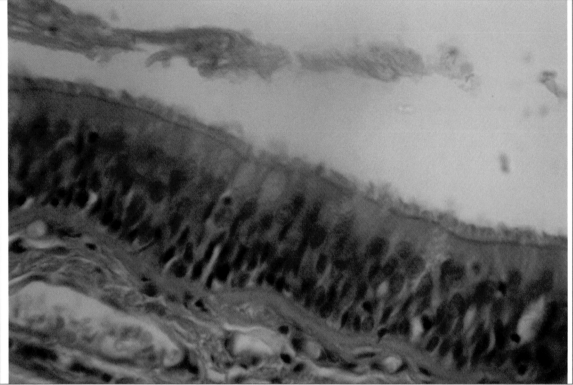

Determination of vital capacity

A spirometer (Figure 6.4) is an instrument used to measure a variety of pulmonary functions:

Tidal volume: the amount of air that is inhaled and exhaled in a relaxed manner;

Inspiratory reserve: the extra amount of air that can be inhaled during forced inhalation;

Expiratory reserve: the extra amount of air that can be exhaled during forced exhalation;

Vital capacity: the total amount of air that can be inhaled and then breathed out. It equals the sum of tidal volume, inspiratory reserve and expiratory reserve as shown in Figure 6.5.

Even after exhaling as much as you can, there is still some air left in your nasal cavity, nasopharynx, trachea, and bronchi. That additional space, which cannot be measured with a spirometer, is called the **residual**.

Figure 6.4. Spirometer apparatus.

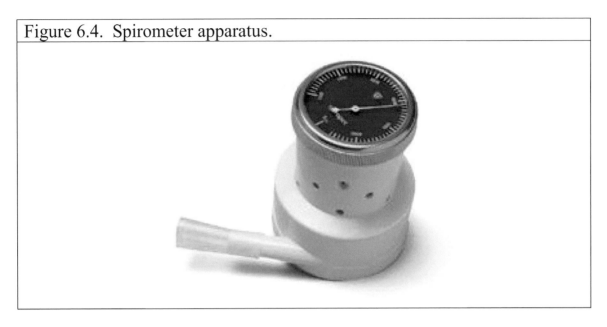

Figure 6.5. Components of lung volume. From: http://members.aol.com/Cappuccinno21/HWAns/HWPicsA/lungvol1.gif

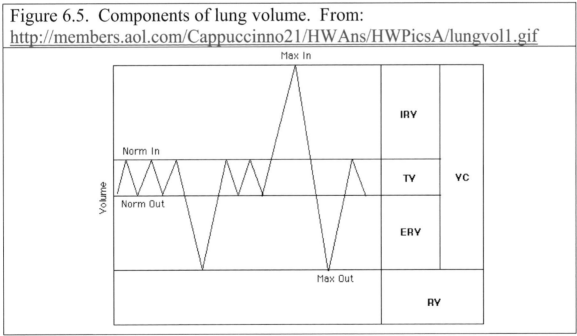

⌘Attach a mouthpiece to the spirometer, and adjust the indicator to "0 liters". Take and exhale 2 deep breaths, and then with the 3rd breath, close your nose and exhale the air into the spirometer. Record your gender, age, height (in inches), and your measurement of vital capacity (in liters). Then indicate, with a "y", or "n", whether you consider yourself athletic (if you play intercollegiate sports or routinely play intramural sports, you are considered athletic for this study,) and whether you smoke cigarettes.

What is your vital capacity?

Complete Table 6.1 with the class data.

Table 6.1. Class data of vital capacity.						
Subject No.	Gender	Age	Height (in inches)	Vital capacity (liters)	Are you athletic? (Y/N)	Do you smoke cigarettes? (Y/N)
1						
2						
3						
4						
5						
6						
7						
8						
9						
10						
11						
12						
13						
14						
15						
16						
17						
18						
19						
20						

Use Microsoft Excel to construct a X-Y scatterplot. Then use the Insert Trendline option, and other Options, add the equation and R-squared value for the line. The R-squared value is a measure of how close the actual points are to the calculated line. R-square can range from +1 to -1. A value of +1 indicates a perfect direct relation between the two variables that are being plotted, while a value of -1 indicates a perfect inverse relation. A value of

"0" indicates absolutely no relation between the two variables. Practically speaking, a value exceeding .75 indicates a strong correlation between the two variables.

✠Add the class data to the vital capacity database (VitalCapacity.xls) which is on the mainframe computer. Calculate the average vital capacity for each of the groups of students indicated in Table 6.2, and then use Microsoft Excel to construct a histogram.

Table 6.2. Average vital capacity for students, according to gender, athleticism, and smoking habits.		
Student group	**Average vital capacity**	**Number in group**
Female, non-athletic, non-smoking		
Female, non-athletic, smoking		
Female, athletic, non-smoking		
Female, athletic, smoking		
Male, non-athletic, non-smoking		
Male, non-athletic, smoking		
Male, athletic, non-smoking		
Male, athletic, smoking		

✠Use the data in Table 6.2 to conduct a Chi-square goodness-of-fit test to determine whether the differences in these average vital capacity values are significant. Do the calculations in Table 6.3.

Student group	Average vital capacity (Observed)	Expected value of vital capacity (equals sum of all observed values divided by 8)	(Observed - Expected)	(Observed - Expected) ^{2}	$(O-E)^2 / E$
Female, non-athletic, non-smoking					
Female, non-athletic, smoking					
Female, athletic, non-smoking					
Female, athletic, smoking					
Male, non-athletic, non-smoking					
Male, non-athletic, smoking					
Male, athletic, non-smoking					
Male, athletic, smoking					
				Chi-square =	
				Degrees of freedom (d.f.) =	7

Table 6.2. Average vital capacity for students, according to gender, athleticism, and smoking habits.

✠Use Table 6.3 to determine the critical value for Chi-square at .95 (i.e. 95% probability) at 7 degrees of freedom. If the calculated value is to the **right** of the critical value, that means that there is only a 5% probability that these results could have come about randomly. If that is the case, then there is only a 5% likelihood that the differences in vital capacity are insignificant.

Table 6.3. Chi-square distribution table.

The Chi-Square (χ^2) Distribution

Area to the Right of the Critical Value

Degrees of freedom	0.995	0.99	0.975	0.95	0.90	0.10	0.05	0.025	0.01	0.005
1	—	—	0.001	0.004	0.016	2.706	3.841	5.034	6.635	7.879
2	0.010	0.020	0.051	0.103	0.211	4.605	5.991	7.378	9.210	10.597
3	0.072	0.115	0.216	0.352	0.584	6.251	7.815	9.348	11.345	12.838
4	0.207	0.297	0.484	0.711	1.064	7.779	9.488	11.143	13.277	14.860
5	0.412	0.554	0.831	1.145	1.610	9.236	11.071	12.833	15.086	16.750
6	0.676	0.872	1.237	1.635	2.204	10.645	12.592	14.449	16.812	18.548
7	0.989	1.239	1.690	2.167	2.833	12.017	14.067	16.013	18.475	20.278
8	1.344	1.646	2.180	2.733	3.490	13.362	15.507	17.535	20.090	21.955
9	1.735	2.088	2.700	3.325	4.168	14.684	16.919	19.023	21.666	23.589
10	2.156	2.558	3.247	3.940	4.865	15.987	18.307	20.483	23.209	25.188
11	2.603	3.053	3.816	4.575	5.578	17.275	19.675	21.920	24.725	26.757
12	3.074	3.571	4.404	5.226	6.304	18.549	21.026	23.337	26.217	28.299
13	3.565	4.107	5.009	5.892	7.042	19.812	22.362	24.736	27.688	29.819
14	4.075	4.660	5.629	6.571	7.790	21.064	23.685	26.119	29.141	31.319
15	4.601	5.229	6.262	7.261	8.547	22.307	24.996	27.488	30.578	32.801
16	5.142	5.812	6.908	7.962	9.312	23.542	26.296	28.845	32.000	34.267
17	5.697	6.408	7.564	8.672	10.085	24.769	27.587	30.191	33.409	35.718
18	6.265	7.015	8.231	9.390	10.865	25.989	28.869	31.526	34.805	37.156
19	6.844	7.633	8.907	10.117	11.651	27.204	30.144	32.852	36.191	38.582
20	7.434	8.260	9.591	10.851	12.443	28.412	31.410	34.170	37.566	39.997
21	8.034	8.897	10.283	11.591	13.240	29.615	32.671	35.479	38.932	41.401
22	8.643	9.542	10.982	12.338	14.042	30.813	33.924	36.781	40.289	42.796
23	9.260	10.196	11.689	13.091	14.848	32.007	35.172	38.076	41.638	44.181
24	9.886	10.856	12.401	13.848	15.659	33.196	36.415	39.364	42.980	45.559
25	10.520	11.524	13.120	14.611	16.473	34.382	37.652	40.646	44.314	46.928
26	11.160	12.198	13.844	15.379	17.292	35.563	38.885	41.923	45.642	48.290
27	11.808	12.879	14.573	16.151	18.114	36.741	40.113	43.194	46.963	49.645
28	12.461	13.565	15.308	16.928	18.939	37.916	41.337	44.461	48.278	50.993
29	13.121	14.257	16.047	17.708	19.768	39.087	42.557	45.722	49.588	52.336
30	13.787	14.954	16.791	18.493	20.599	40.256	43.773	46.979	50.892	53.672
40	20.707	22.164	24.433	26.509	29.051	51.805	55.758	59.342	63.691	66.766
50	27.991	29.707	32.357	34.764	37.689	63.167	67.505	71.420	76.154	79.490
60	35.534	37.485	40.482	43.188	46.459	74.397	79.082	83.298	88.379	91.952
70	43.275	45.442	48.758	51.739	55.329	85.527	90.531	95.023	100.425	104.215
80	51.172	53.540	57.153	60.391	64.278	96.578	101.879	106.629	112.329	116.321
90	59.196	61.754	65.647	69.126	73.291	107.565	113.145	118.136	124.116	128.299
100	67.328	70.065	74.222	77.929	82.358	118.498	124.342	129.561	135.807	140.169

Donald B. Owen, *Handbook of Statistical Tables*, U.S. Department of Energy (Reading, Mass.: Addison-Wesley. 1962). Reprinted with permission of the publisher.

Exercise 7: Determination of tar content of selected cigarettes and effects of long-term exposure

Equipment
Dissecting microscopes
Vacuum pump
Cigarette extraction apparatus
Electronic balance, capable of measuring milligrams

Materials
Low-tar and high-tar varieties of cigarettes
Preserved slides
       Lung cross-section with carbon particles
       Lung cross-section with emphysema
       Lung cross-section with cancer

Background
Epidemiological data indicate that cigarette smoking accounts for almost 0.5 million deaths per year in the United States (Figure 7.1).

Furthermore, over 4,000 compounds have been isolated from cigarette smoke. The major components, and their short-term and long-term effects are shown in Table 7.1.

In this experiment, we will be extracting and measuring the amount of tar present in selected cigarettes, including at least one brand that is marketed as a low-tar cigarette and a second that is marketed as a regular cigarette. We will also be looking at histological samples of tissues which show two of the common consequences of long-term smoking -- emphysema and cancer.

Figure 7.1. U.S. deaths caused by tobacco as compared to other drugs of abuse.

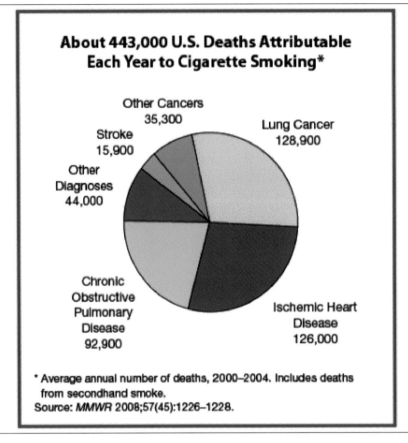

Table 7.1. Short-term and long- term effects of chemical components in cigarette smoke.

Component	Short-term effects	Long-term effects
Nicotine (Range: 0.1-1.6 mg)	1. Increase in blood pressure 2. Fat mobilization 3. Paralysis of ciliated epithelium 4. Increase in pulse	1. Atherosclerosis 2. Persistence of dust particles in lung 3. Decrease in heart reserve
Tar - particulate matter (Range: 1-80 mg)		Replacement of epithelial tissue with scar tissue
Carbon monoxide	Headache, drowsiness, inability to concentrate.	?

Carcinogens		Onset of malignancies.
• Benzopyrene • Chrysene • Nitrosamines • Nickel compounds		

✖Measuring tar content of selected cigarettes

The combustion of the cigarette should be performed in a hood, while the weighing can be done on one of the benches.

1) Attach the vacuum pump to the tar extraction apparatus;
2) For each cigarette measured:
    a) carefully weigh a filter, and record the weight in the appropriate space in Table 7.3;
    b) attach the filter to the apparatus. Make sure there are no tears on the filter;
    c) Light the cigarette with a match or candle and then immediately turn on the vacuum pump. It will take approximately 10 seconds for the cigarette to be completely burned. You will notice smoke in the tar extraction apparatus;
    d) When the flames are within 2 mm of the filter, remove the cigarette with tweezers or forceps, and allow the vacuum pump to run until the smoke has been adsorbed onto the filter;
    e) Turn off the vacuum pump, take apart the extraction apparatus, and let the filter dry for 10 minutes before weighing again. Record the weight in the appropriate space in Table 7.1;
3) Use at least 5 low-tar and 5 high-tar cigarettes, preferably the same brand.
4) If only two varieties of cigarettes were tested, then the data can be analyzed with a Student's t-test. If more than 2 varieties of cigarettes were tested, then the data must be analyzed with an analysis of variance (ANOVA).

Table 7.3. Results of tar-determination test, comparing low-tar and high-tar brands. (Write down the brands that were used.)					
Low-tar brand:			High-tar brand:		
Initial weight of filter (mg)	Final weight of filter (mg)	Difference (= tar content)	Initial weight of filter (mg)	Final weight of filter (mg)	Difference (= tar content)
Average tar content:			**Average tar content:**		
Degrees of freedom (d.f.) =					
Critical value of 't' at 95% confidence level					
Calculated value of 't'					

✻Effects of long-term exposure

Chronic cigarette smoking often sets into motion a series of changes which eventually are irreversible. First, there is the deposition of carbon particles, which will induce a cellular response in an attempt to "wall off" or isolate the offending deposits of carbon particles. This will often result in the replacement of health epithelial tissue with scar tissue. Alveolar spaces are thereby obliterated by either the deposition of carbon particles or the cellular response. (Figure 7.2)

Figure 7.2. Lung tissue showing the accumulation of carbon particles.

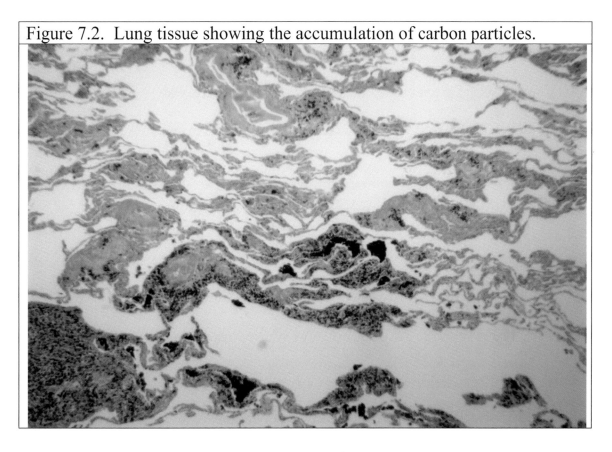

In the case of emphysema, the integrity of the supporting tissues in the bronchioles is compromised, so that they collapse every time that the individual attempts to inhale. As a result, air spaces in the lungs become grossly enlarged , and the person cannot take a deep breath. People with emphysema report that they feel like they are constantly trying to breathe through a straw (Figure 7.3)

Figure 7.3. Cross-section of a lung showing characteristics of emphysema.

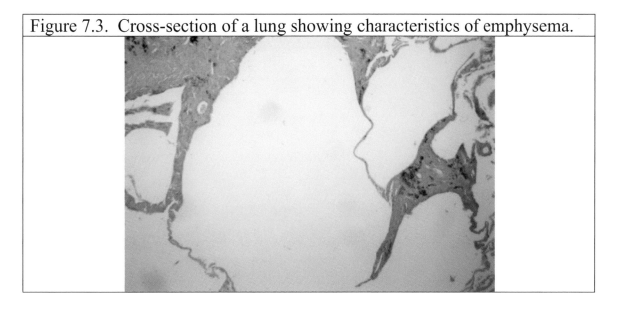

In lung cancer, carcinogenic chemicals induce mutations in oncogenes and/or tumor suppressor genes. These progressive, cumulative changes will eventually induce a cell, or some cells, to lose their ability to undergo contact inhibition and they will therefore start to grow into a tumor (Figure 7.4).

Figure 7.4. Cancers involve progressive changes in tissues. From http://bio3400.nicerweb.com/Locked/media/ch18/18_03-cervical_cancer.jpg

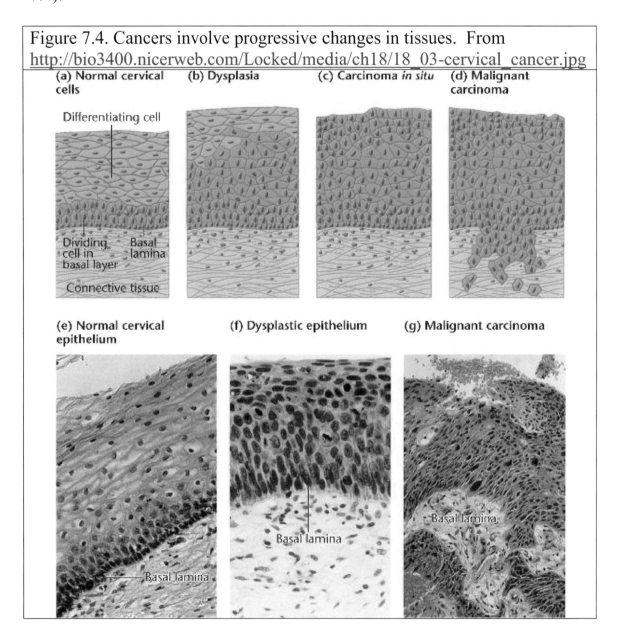

The differences in gross anatomy between a healthy lung and a diseased lung can be quite dramatic (Figure 7.5).

Figure 7.5. Contrast between a healthy lung and a lung with cancer.	
a. A healthy lung. Note that it is fairly uniform in color and texture. From: http://www.lungusa.org/learn/art/clean_lung_hr.jpg	b. This lung with a prominent beige tumor is darkened from cigarette tar. From: http://www.abc.net.au/science/news/img/lungcan.gif
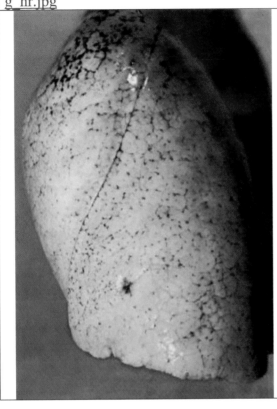	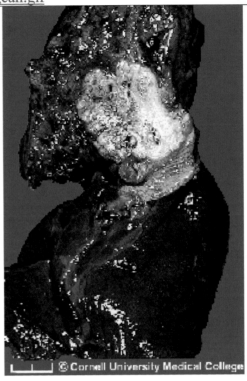

Exercise 8. Circulatory System and the effects of caffeine on the cardiovascular system.

(Adapted from Lab Topic 28: Investigating Circulatory Systems, in Warren Dolphin's Biological Investigations, edition 6.)

Equipment
Compound microscope
Dissecting microscope
Blood pressure kits, including a sphygmomanometer and a stethoscope
Electrocardiogram apparatus

Supplies
Prepared slides
Human blood smear stained with Giemsa stain
Artery & vein c.s.
Cardiac muscle l.s.
Sheep heart
Human heart models
Fetal pigs
Dissecting tools and sets
Isopropyl alcohol and electrode cream, if necessary, for electrocardiogram
1 Stender dish with 500 ml aquarium water and .1 gram MS-222 (methyl tricaine sulfonate), labeled "Anesthesia"
1 Stender dish with 500 ml aquarium water, labeled "Recovery"
goldfish
cotton balls
microscope slides & coverslips
1 bottle of NoDoz (or equivalent) caffeine tablets
drinking cups

Objectives

1) To dissect the heart and major vessels of the mammalian circulatory system;
2) To observe blood flow in capillary beds of a fish;
3) To determine the short-term effects of caffeine on cardiovascular function, particularly pulse, blood pressure, and electrocardiogram pattern.

Background

Invertebrates, such as cnidarians, flatworms, and roundworms, lack circulatory systems. Simple diffusion is sufficient for the necessary exchanges of respiratory gases, waste products, and nutrients. Larger, more complex animals require a circulatory system to supply the needs of their tissues.

The circulatory system consists of a special internal body fluid called **blood** or **hemolymph**, a pumping system, and a vascular system consisting of vessels for moving the blood rapidly from one location to another within an animal. The circulating fluid often contains a respiratory pigment, a pigment that aids in transporting oxygen and carbon dioxide between the tissues and the respiratory surface. Hemocyanin and hemoglobin (which in mammals occurs in red blood cells) are common pigments. Blood also contains cells or proteins that protect against invasion by microorganisms and proteins that are involved in clotting, the sealing of leaks. The blood-vessel system often anatomical provisions so that the blood is brought into close contact with three other physiologic systems: lung or gill, where gas exchange occurs; excretory, where waste, salt, and water exchange occur; and digestive, where nutrients are absorbed.

In the closed circulatory systems of vertebrates, the flow is blood is always within blood vessels. The arterial system is connected to the venous system by means of capillaries which have very thin walls only one cell thick. Blood entering the capillaries is under relatively high pressure, and part of the fluid portion is filtered through the capillary walls, entering the tissue spaces. On the venous side of the capillary bed, most of this fluid flows back into the capillaries due to osmotic force. Gaseous, waste, and nutrient exchanges between the blood and tissues occurs by way of this fluid exchange as well as by diffusion. Blood flow in each capillary bed is regulated by the opening and closing of a **precapillary sphincter**.

The capillary bed, and only the capillary bed, is the functional site of the closed circulatory system where all exchanges take place. The **lymphatic system** consists of small open-ended lymphatic capillaries that conduct fluid into larger lymphatic ducts. Fluid that does not return to the blood capillaries enters the lymphatic capillaries. This fluid is collected in lymphatic ducts and is returned to the venous system near the heart.

Mammalian Circulatory System

If your fetal pig has not previously been opened, make a series of cuts as diagramed in Figure 3.1. If you have followed the lab sequenc3 in this manual, complete the opening as follows:

1) Make a longitudinal cut 1 cm to the left of the sternum from the lower ribs to the region of the forelimbs and parallel to the previous cut. Sever all ribs;

2) List up the center section of tissue and cut any tissues adhering underneath. A transverse cut at the anterior end will detach this center piece, which should be discarded;

3) The heart and lungs will be easier to observe if the diaphragm is cut away from the rib cage on the animal's left side only. Cut close to the ribs;

4) In the region of the throat, remove the thymus glands, thyroid, and muscle bands, but do not cut or tear any major blood vessels.

The Heart and Its Vessels

Find th ehart encased in the **pericardial sac**. Remove the sac and identify the four heart chambers. The paired **atria**, thin-walled, distensible sacs, collect blood as it returns to the heart. The two **ventricles** are the large, muscular pumping chambers of the heart.

Blood returning from the systemic circulation enters the right atrium from the cranial and caudal **vena cavae**. After passing into the right ventricle, it is pumped to the lungs through the **pulmonary trunk**, which divides into the left and right **pulmonary arteries**. The trunk is visible passing from bottom right to upper left over the front of the heart and passing between the two atria. Trace the pulmonary arteries to the lungs.

Following gas exchange in the capillaries of the lungs, blood collects in the **pulmonary veins** and flows to the left atrium. These veins enter on the dorsal side of the heart and will be difficult to find. If you remove some of the lung tissue from the left side, you may be able to locate these vessels.

✠From the left ventricle, blood is pumped at high pressure through the **aorta** to the systemic circulation. Find the aorta. It will be partially covered by the pulmonary trunk but can be identified as the major vessel that curves 180° to the left, forming the **aortic arch**. In the mammalian fetus, the pulmonary trunk and aorta are connected by a short, shunting vessel, the **ductus arteriosus**. Find this vessel in your animal. During the intrauterine life, when the lungs are not functional, blood entering the pulmonary circuit does not pass to the lungs. Instead, it is shunted to the aorta. At birth, the shunting vessel constricts so that blood enters the lungs. The constricted vessel fills with connective tissue to become a solid cord seen in adults as the arterial ligament.

Vessels Cranial to the Heart

Veins

Because the venous system is generally ventral to the arterial system, it will be studied first in the congested region of the heart.

Trace the **cranial vena cava** forward from the heart to where it is formed by the union of the two very short **brachiocephalic veins**. Each of these in tern is formed by the union of the five major veins: the **internal** and **external jugular veins**, which drain the head and neck; the **cephalic vein**, which lies beneath the skin anterior to the upper forelimb and typically enters at the base of the external jugular; the **scubscapular vein** from the dorsal aspect of the shoulder; and the **subclavian vein** from the shoulder and forelimb. As the latter passes into the forelimb, it is known as the **axillary vein** in the armpit and the **brachial vein** in the upper forelimb. Caudal to the union of the brachiocephalic veins, find the pair of **internal thoracic veins** entering the ventral surface of the vena cava. They extend along the sternum and drain the chest wall. These veins were most likely cut when you opened the animal.

Arteries

Find the aortic arch and trace it back to the heart. Note the several arteries that branch off to supply the anterior region of the animal.

Find the small **coronary arteries** that arise from the base of the aorta behind the pulmonary trunk. They pass to the groove between the ventricles on the ventral surface of the heart and branch to supply the muscles of the heart. The first major artery to branch from the aorta is the **brachiocephalic artery.** It gives rise to the two **carotid arteries,** which pass anteriorly to supply the head, and the **right subclavian artery,** which passes to the right forelimb. Just to the left of the brachiocephalic artery, find the **left subclavian artery** arising as a separate branch from the aortic arch. Blood in this vessel goes to which region of the body?

Once the aorta runs posteriorward along the dorsal wall of the thorax, it gives rise to intercostal arteries, which supply the walls of the chest. The aorta then passes through the diaphragm to become the **abdominal aorta.**

Return to the left subclavian artery and trace it into the forelimb, removing skin and separating muscles as necessary. In the armpit it is known as the **axilary artery,** and in the upper forelimb as the **brachial artery.** The **subscapular artery** branches from the axillary artery and supplies the shoulder muscles. The brachial artery divides in the lower forelimb to give rise to the **radial** and **ulnar arteries.**

Vessels Caudal to the Heart

Veins

If the heart is lifted and tilted forward the **caudal vena cava** can be viewed at the point where it enters the right atrium. As this vein is traced caudally, several veins will be found flowing into it. After it passes through the diaphragm, the paired **hepatic veins** and single **umbilical vein** enter first. The umbilical vein carries oxygenated, nutrient-laden blood from the placenta. This vein passes through the liver where it is known as the ductus venosus.

While studying the major veins of the body cavity, note the **hepatic portal system**. It consists of veins that flow from the digestive tract and spleen to the liver where they again divide into a system of capillaries. The term **portal system** refers to a system of veins that arises from a capillary bed and carries blood to a second location. There the veins again fan out to form capillaries, which again flow into veins before the blood returns to the heart.

The hepatic portal system can be traced partway from the liver to the digestive system. The **portal vein** runs next to the common bile duct in the **hepatoduodenal** ligament under the lobes of the liver. Nutrient-laden blood flows up this vein to the liver where exchange occurs between the blood and the liver across the walls of the portal system capillaries. The blood then flows into the **hepatic veins,** which enter the caudal vena cava.

Follow the vena cava caudally to where the **renal veins** enter from the kidneys. In the male, the **spermatic veins,** and in the female, the **ovarian veins**, enter next. On the left side, these veins may enter the renal vein first. Below the kidneys, the vena cava splits into the **internal** and **external iliac veins** and the **median sacral vein**, a small vein that comes from the tail. The external iliac veins collect blood from the **femoral veins** in the hind legs, whereas the internal iliacs collect blood from the pelvic area.

Arteries

After the aorta enters the abdominal cavity, a large single **coeliac artery** arises from it at the cranial end of the kidneys. You will have to remove some connective tissues to obtain a full view of this artery. The coeliac artery eventually divides into three arteries supplying the stomach, spleen, and liver. The **mesenteric artery** next arises from the aorta and supplies the pancreas, small intestine, and large intestine. The **renal arteries** are short, paired arteries supplying the kidneys. The next large arteries arising from the aorta are the **external iliacs**, which supply the hind legs with an **iliolumbar** branch to the lower back. In the fetus the **umbilical arteries** branch from the caudal end of the abdominal aorta and pass out through the umbilical cord. They form a capillary bed in the placenta for nutrient, gas, and waste exchange with the maternal circulatory system.

Internal Heart Structure

Study the orientation of the heart so that you can later identify it in isolation. Now, free the heart from the body by cutting through all the vessels holding it in place. Be careful to leave enough of each vessel so that they can be identified in the isolated heart. Alternative to removing the fetal pig's heart, your instructor may have a demonstration dissection of a beef or sheep heart for your study. Whatever specimen you are using, orient yourself by identifying the **aorta, pulmonary artery, pulmonary vein,** and **vena cava.**

✠Place the heart in your dissecting pan, ventral side up. Make a razor cut along the pulmonary trunk down through the right ventricle. Spread the tissue open, pin it down, and remove the latex. You may wish to use a dissecting microscope to observe the open ventricle.

✠Identify the **semilunar valves** at the junction of the artery and ventricle. Consider how these valves work. The open flaps face into the **pulmonary trunk**, and any backflow in the pulmonary trunk fills the valve flaps with blood and closes the valve. You may have to add water to the heart to float the valve flaps so that you can see them.

✠Now cut through the right atrium and remove the latex and coagulated blood. The **tricuspid valves** are between the atrium and ventricle. These valves also work on the backflow principles, allowing blood to flow only one way from the atrium into the ventricle. In the ventricle, fine fibers called **chordate tendinae** are attached to the valve flaps. These cords prevent the flaps from "blowing back" from high pressures developed when the ventricle contracts.

✠Cut into the left atrium and ventricle as you did on the right side. Identify the **bicuspid** or **mitral valve** with its associated chordae tendinae, between the atrium and ventricle. After cutting into the ventricle and cleaning it, find the **aortic semilunar valve.** Blood flow through the human heart is shown in Figure 8.2.

Figure 8.2. Blood flow through the human heart. From: http://www.gsu.edu/~bioasx/humanheart.gif

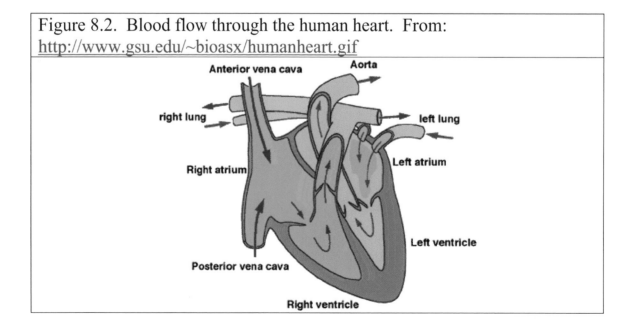

86

Clean your dissecting instruments and tray and return your fetal pig to the storage area.

Histology of the circulatory system

Vessels

✠Obtain prepared slides of cross sections of arteries and veins and observe them under low power with a compound microscope. Note that arteries have thicker walls than veins. Most of the difference in thickness is due to the increased amounts of muscle and connective tissue in the artery. Since arteries carry blood from the heart, they operate under relatively high pressure (average 120 mm of mercury equivalent). Veins experience only a tenth as much pressure.

Blood flows through veins because skeletal muscles press on them and move the blood along. **Valves** in the veins prevent backflow and make the passgae of blood unidirectional.

Blood

Human blood consists of 55% plasma and 45% cells by volume. Plasma is the fluid portion of the blood containing dissolved proteins, salts, nutrients, and waste products. Several different types of cells and cell fragments are contained in blood. Get a prepared slide of a Wright-stained human blood smear and look at it with your compound microscope under medium power (Figure 8.3).

Figure 8.3. Human blood smear.

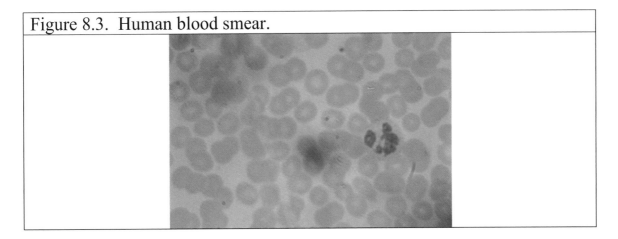

By far the most common (about 95% of the cells) are erythrocytes (red blood cells) which are red because they contain hemoglobin, an oxygen-carrying protein. The other 55 are collectively called **leukocytes** (white blood cells) and **platelets** that are important in blood clotting. There are several types of leukocytes, including neutrophils, basophils, eosinophils, lymphocytes and monocytes.

Neutrophils leave the blood early in the inflammation process and become phagocytic cells consuming cell debris and bacteria. **Lymphocytes** are responsible for our ability to mount a specific immune system. Some are involved in cellular immunity (T-lymphocytes), while others (B-lymphocytes) secrete antibodies that neutralize foreign proteins and other macromolecules.

Cardiac muscle

The heart consists of cardiac muscle tissue. These muscle cells are uninucleated and partially striated. Adjacent cells are connected by intercalated disks, which allow cytoplasm, and more importantly, electrochemical impulses, to flow freely from one cell to another (Figure 8.4).

Figure 8.4. Cardiac tissue.

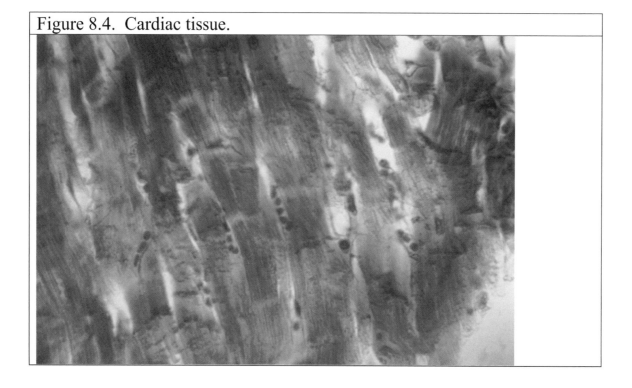

Circulation in capillaries

To observe circulation in capillaries, net a small fish or tadpole from an aquarium and gently drop it into a Stender dish with MS-222, an anesthetic for aquatic organisms. When the fish does not respond to gentle prodding with a pencil or glass rod, wrap it in dripping wet common, being careful not to cover the head or the tail. Lay the wrapped fish in an open Petri dish. Place a few drops of aquarium water on the tail and add a coverslip over the tail.

Place the dish on a compound microscope stage and observe the tail under scanning power. Sketch your observations, answering the following questions:

1) Can you identify capillaries, venules, and arterioles?

2) Is blood flow faster in certain vessels compared to others?

Is blood flow continuous in all vessels? What might control this?

Dilute solutions of nicotine, caffeine, and adrenalin are available in the lab in dropper bottles. You can see the effects of these chemicals by placing a drop on the tail.

✠Effects of Caffeine on Pulse and Blood Pressure

Caffeine is described as "an effective psychostimulant [whose behavioral effects] include increased alertness, a faster and clearer flow of thought, wakefulness, and restlessness (Julien, 1998). It also has a "slight stimulant action on the heart, increasing cardiac contractility and output and dilating the coronary arteries".

The purpose of this experiment is to explore the effects of caffeine on cardiovascular output, as measured by pulse and blood pressure.

If you are willing to be research subjects in this experiment, please read the **informed consent form**, detach it, sign your name at the appropriate place, and return the form to your instructor.

In preparation for conducting this experiment, you are asked to refrain from drinking any caffeinated drink after midnight Sunday night. At the beginning of class, we will take baseline pulse and blood pressure data, using a stethoscope and a sphygmomanometer. The cuff of the sphygmomanometer will be placed on one of your arms above the elbow joint. A stethoscope will then be placed underneath the cuff and the investigator will then pump air into the cuff to reach a pressure of 200 mm mercury, which should be enough to close the brachial artery. By slowly releasing air from the sphygmomanometer, the investigator will be able to measure the pressure at which pressure sounds are detectable through the stethoscope (corresponding to your systolic pressure) and at which pressure those sounds stop (corresponding to your diastolic pressure). You may feel temporary discomfort by the pressure of the cuff on your arm. As the investogator releases air from the cuff, the discomfort will stop. Pulse (heartbeats/minute) can be determined by feeling the number of beats within 15 seconds and then multiplying that amount by 4.

After taking baseline measurements, you will be asked to ingest one NoDoz tablet, which has 200 milligrams of caffeine (equivalent to two 5-ounce cups of coffee) with enough water to wash it down. The box states the following warnings:

From the Directions of Use on the box:

Keep this and all other medications out of the reach of children. In case of accidental overdose, seek professional assistance or contact a poison control center immediately. As with any drug, if you are pregnant or nursing a baby, seek the advice of a health professional before using this product. Do not give to children under 12 years of age. For occasional use only. Not intended for use as a substitute for sleep. If fatigue or drowsiness persists or continues to occur, consult a doctor. The recommended dose of this product contains as much caffeine as a cup of coffee *(N.B.: Note the discrepancy between what the box says and what the textbook says.)* Limit the use of caffeine-containing medications, food, or beverages while taking this product because too much caffeine may cause nervousness, irritability, sleeplessness and, occasionally, rapid heart beat.

Do not participate as a research subject if you are allergic to the following additional ingredients found in each tablet: Benzoic acid, Carnauba wax, corn starch, FD&C Blue dye No. 1, hydroxypropyl methylcellulose, microcystalline cellulose, mineral oil, Polysorbate 20, povidone, propylene glycol, simethicone emulsion, sorbitan monolaurate, stearic acid, sucrose, titanium dioxide.

Procedure

1) Fill in Table 8.1, which summarizes your daily caffeine intake. Calculate the daily caffeine intake for the class;
2) Measure initial systolic pressure, diastolic pressure, and pulse, and enter these data in the appropriate cells of Table 8.2;
3) Ingest one NoDoz (or equivalent) tablet with sufficient water to be able to swallow it;
4) One hour after ingesting the NoDoz tablet, measure final systolic pressure, diastolic pressure, and pulse. Enter your personal data in the appropriate cells of Table 8.2;
5) Compile the class data into Table 8.3, and then add the class data to the pre-existing Excel file of data from previous classes and determine average values for initial and final systolic pressure, initial and final diastolic pressure, and initial and final pulse
6) Construct 1 histogram for each of the following:
   a) Initial and final systolic pressure
   b) Initial and final diastolic pressure
   c) Initial and final pulse
   d) Frequency histogram for daily caffeine intake

7) Determine with Student t-tests whether the differences between initial and final values for systolic pressure, diastolic pressure, and pulse are significant;
8) Use the histograms as Figures in your laboratory report.

References Cited:

Julien, Robert M. 1998. A Primer of Drug Action. W.H. Freeman and Company, New York.

RELEASE OF ALL CLAIMS
Effects of caffeine on cardiovascular function

Release executed on this _____ day of _____,, _____ (Year).

    As a lawful consideration for being permitted by Christian Brothers University to participate in the below-described activity, I, the undersigned, hereby agree that I, my family, heirs, distributees, guardians, legal representatives and assigns:

    1. Will not make a claim against, sue, attach the property of, or prosecute Christian Brothers University or any of its employees, agents, or instructors, (hereafter "CBU"), for any injury or damage that may occur during my voluntary participation in such activity, even though liability may arise out of the negligence or carelessness on the part of the persons or entities mentioned above.

    2. Will save and hold harmless CBU from any claims by me, or my family, heirs, distributees, guardians, legal representatives or assigns arising out of my voluntary participation in such activity.

    I hereby personally assume all risks of loss, damage, or injury that I may sustain in connection therewith, whether those risks are foreseen or unforeseen, and whether by negligence or not.

    I further release all agents and employees from any claim whatsoever on account of first aid, treatment or service rendered me during my participation in such activity.

    I further state that I am of lawful age and legally competent to sign this release. This release contains the entire agreement between the parties hereto and the terms of this release are contractual and not a mere recital.

    I HAVE CAREFULLY READ THIS AGREEMENT AND FULLY UNDERSTAND ITS CONTENTS. I AM AWARE THAT THIS IS A RELEASE OF LIABILITY AND A CONTRACT BETWEEN MYSELF AND CHRISTIAN BROTHERS UNIVERSITY AND/OR ITS AGENTS AND EMPLOYEES, AND I VOLUNTARILY SIGN THIS WAIVER AND RELEASE.

DESCRIPTION OF ACTIVITY:
This experiment involves a study of caffeine on cardiovascular function. As part of this experiment, you will do the following:
1) Determine your daily caffeine intake;
2) Determine initial systolic blood pressure, diastolic blood pressure, and pulse;
3) Ingest one NoDoz (or equivalent) tablet, containing 200 mg of caffeine;
4) One hour after ingesting the caffeine, you will determine final systolic blood pressure, diastolic blood pressure and pulse.

As a result of caffeine ingestion, you may notice an increase in alertness, irritability, and cardiac responses. These reactions are temporary, and should subside as the caffeine is metabolized, usually within 4 hours.

Signature: _____

Table 8.1. Estimate of Daily Intake. Fill in the appropriate blanks by indicating how much of each type of beverage or food you consume each day.			
Item	**Average content (milligrams)**	**# or amount ingested per day**	**Total (= Average content * # or amount)**
Coffee (5-oz cup)	100		
Tea (5-oz cup)	50		
Cocoa (5-oz cup)	5		
Chocolate (semisweet)	25		
Chocolate milk	5		
Cola drinks (12 oz cans)	40		
Cola drinks (20 oz bottles)	67		
OTC stimulants (NoDoz, Vivarin)	200		
OTC analgesics (Excedrin)	65		
OTC cold remedies (Triaminicin)	30		
OTC diuretics (Aqua-ban)	100		

Note: OTC = Over the counter   **Total:**

Table 8.2. Personal data regarding the effects of caffeine on cardiovascular function.			
When is the last time you ingested a beverage or food with caffeine?			
What is your gender?			
What is your age?			
What is your weight?			
Time	**Pulse (beats per minute)**	**Systolic Pressure (mm Hg)**	**Diastolic pressure (mm Hg)**
Initial (Pre-ingestion)			
Final (One hour after ingestion)			

No.	Gender	Age	Weight (lbs)	Pulse (beats/ minute)	Initial Systolic Pressure	Final Systolic Pressure	Initial Diastolic Pressure	Final Diastolic Pressure
1								
2								
3								
4								
5								
6								
7								
8								
9								
10								
11								
12								
13								
14								
15								
16								
17								
18								
19								
20								

Figure 8.3. Class data for caffeine experiment.

9.   Effects of selected drugs on *Daphnia magna*, a microcrustacean

Daphnia culture protocol, from Carolina

Daphnia pulex
Culture Instructions

Upon arrival, pour the gallon of culture water into the plastic aquarium. Remove the lid from the jar of daphnia and slowly immerse the jar in the culture water until it is completely filled. While the jar is submerged, pour the culture of *Daphnia pulex* into the aquarium. Introducing the daphnia into the aquarium in this manner prevents air bubbles from becoming trapped under their carapace. The bubbles might lift them to the surface, possibly killing them.

Next, add $1/16$ teaspoon (a pinch) of Daphnia Food to the aquarium. **Note:** It is very important that no more than $1/16$ tsp of food be added initially. Adding additional food does not increase the reproductive rate of the daphnia but instead causes the bacteria in the aquarium to reproduce much faster than the daphnia can consume them. This sudden increase in bacteria raises the biological oxygen demand in the aquarium and lowers the dissolved oxygen level to the point that the daphnia are displaced.

Once bacteria begin to break down the food pellets and their population increases, the culture will become cloudy or turbid. The daphnia will begin to "graze" upon the bacteria suspended in the water and in 7–10 days will reduce the amount of bacteria. Once the water in the aquarium begins to clear, add another $1/16$ tsp of food to the aquarium. As the population of *Daphnia pulex* in the aquarium increases, they will consume the bacteria at a faster rate. At this point it may become necessary to feed the daphnia more frequently.

Over the life span of the culture, debris will accumulate on the bottom of the aquarium. Do not attempt to remove this debris as it will contain ephippial eggs (fertilized, resting eggs) that will hatch later. Algae may also begin to grow on the sides of the aquarium. Adding several small snails will eliminate any algae buildup and will help to promote a healthier culture of daphnia.

The daphnia need light for a healthy culture, but keep the aquarium out of direct sunlight since the ultraviolet light will have an adverse effect. Ambient light from overhead fluorescent lights will suffice. Maintain at 68–72°F.

Daphnia on Drugs

Adapted from an experiment prepared by Joan K. McLaughlin-Johnston

This laboratory is an enjoyable, safe, hands-on learning experience for students. As a consequence of its adapted cooperative learning format and constructivist procedural setup, it addresses many learning styles and ability levels that may be present in a diverse classroom population.

PROBLEM
How do various chemicals affect the heart rate of a Daphnia?

MATERIALS

microscope	medicine dropper	depression slide
coverslip	paper towel	stopwatch
calculator	Daphnia	coffee solution
tea solution	aspirin solution	cigarette solution
scotch solution	sleeping pill solution	adrenalin solution

PROCEDURE

1. Students should work in pairs. When students have completed reading the written directions, use the medicine dropper to withdraw a Daphnia from the container in which it is stored. Place the Daphnia together

with a few drops of water from the container into a well in the depression slide. Cover the Daphnia with a coverslip.

2. Observe the Daphnia under low power. Locate the heart using the diagram provided. Locate the various organs labeled.

3. Using your stopwatch, record how many heartbeats you can count in one minute. Record the heartrate in your data chart.

4. Repeat for two additional trials. Record your data. Calculate the average heartrate of your daphnia. Record your baseline data in Table 9.1.

5. Use the medicine dropper to transfer two drops of assigned chemical to one edge of the coverslip. Draw the chemical across the specimen by placing a piece of paper towel at the opposite edge of the coverslip.

6. Determine the effects of the chemical on the heartbeat by repeating steps #3 through 5. Record the experimental data in Table 9.2.

7. Repeat steps #3 through 6 with each drug type. (Given sufficient number of *Daphnia*, it may be possible to use a new *Daphnia* for each drug.)

9. Prepare a graph of the data you have collected on the heartbeat both before and after exposure to the chemical assigned.

	1% EtOH		5% EtOH		10% EtOH		20% EtOH		Caffeine		Xanax		Adderall	
	Base-line	Pos t	Base-line	Pos t	Base-line	Pos t	Base-line	Pos t	Base-line	Pos t	Base-line	Pos t	Base-line	Pos t
Rep-licat e #1														
Rep-licat e #2														
Rep-licat e #3														
AVG.														

Table 9.1. Effects of selected drugs on *Daphnia magna*.

OBSERVATIONS

1. Describe the Daphnia's heartrate.

2. What was the Daphnia's average heart rate before exposure to each chemical?

3. What was the Daphnia's average heart rate after exposure to the each chemical?

CONCLUSIONS

1. What effect did the chemical have upon the Daphnia's heart rate?
2. What are the active ingredients in tea, scotch, sleeping pills, cigarette extract, and Adderall?
3. What do we call drugs that cause an increase in heart rate in humans?
4. What do we call drugs that cause a decrease in heart rate in humans?

Exercise 10: Ethanol. I. Effects on humans: Gross anatomy of liver: 1)Normal; 2) Fatty infiltration (alcohol liver disease); 3) Cirrhosis

Equipment
Compound microscope
Dissecting microscope

Supplies
Prepared slides
1) Liver cross-section: Normal
2) Liver cross-section: Fatty infiltration (indicating alcohol liver disease)
3) Liver cross-section: Cirrhosis

Plasticmount of liver with cirrhosis

Background
Ethanol is a non-specific depressant, which acts to facilitate GABA receptors. The effects of ethanol are dose-dependent, with low doses interfering with equilibrium, reflexes, judgement and coordination (Figure 9.1). Extremely high doses can suppress breathing.

Although a small amount of ethanol can be absorbed through the mucous membranes of the mouth, most of it is absorbed in the small intestine. Maximum blood concentrations are reached within 30 minutes of ingestion.

The purpose of this exercise is to examine histological specimens of the liver in progressive states of damage due to ingestion of ethanol.

Figure 9.1. Likelihood of being in a traffic accident as a function of alcohol intake

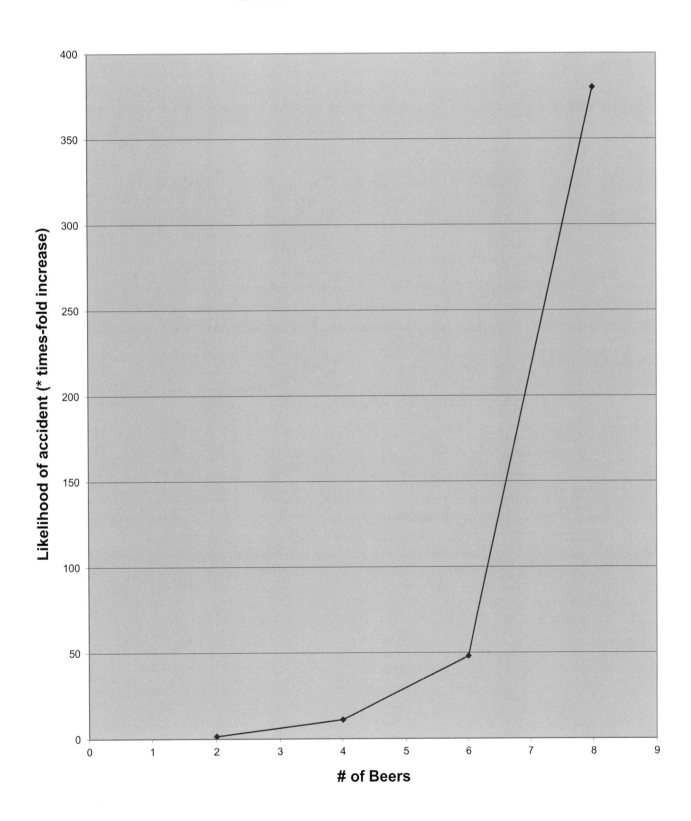

�֎Histology of chronic ethanol ingestion

The hepatic portal system directs all substances absorbed from the small intestine to the liver. The metabolism of ethanol occurs in the liver, whose cells possess a series of enzymes which catalyze the breakdown of ethanol to water and carbon dioxide as final waste products.

The breakdown of ethanol is initiated by ethanol dehydrogenase, which catalyzes the conversion of ethanol to acetaldehyde. The second enzyme in the pathway, acetaldehyde dehydrogenase, converts acetaldehyde to acetic acid.

These metabolic steps occurring in the liver routinely generate large quantities of free radicals, so that chronic ingestion of ethanol is harmful. Initial responses include inflammation and infiltration of fatty tissue, which are indicative of alcoholic liver disease. At this point, the damage caused by the ingestion of alcohol is reversible.

With continued ingestion of ethanol, however, functional liver tissue is replaced by scar tissue, which is incapable of undergoing metabolic breakdown of food or toxins. This condition, called **cirrhosis**, is life-threatening.

✖Work in groups of 3 to examine cross-sections of normal liver tissue, liver tissue showing signs of fatty infiltration, and liver tissue showing cirrhosis. Each group of students should obtain 1 slide each of normal liver tissue, liver tissue showing fatty infiltration, and cirrhotic liver tissue. Scan each of these slides under 40x, and then focus on a small section of each slide under 400x. Draw what you see in the appropriate space in Figure 9.2.

Figure 9.2. Three views of liver tissue.

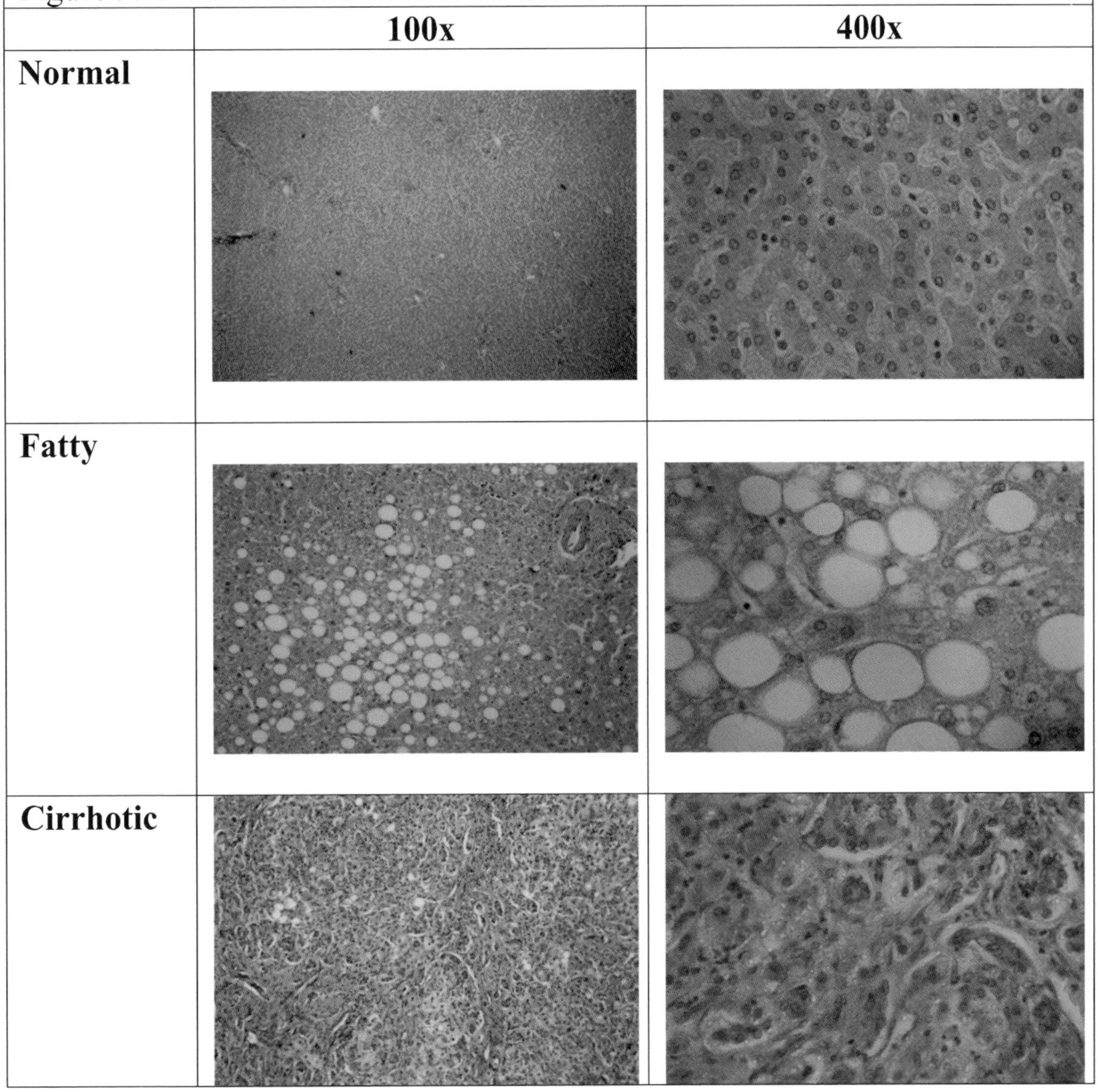

Gross anatomy of cirrhotic liver

Obtain a plasticmount showing a diseased liver. Notice the blackened areas, indicating areas of cirrhosis.

Exercise 11: Ethanol. Effects on *Drosophila melanogaster*: Fecundity and average hatch weights

Equipment
Compound microscope
Dissecting microscope
Analytical balances

Supplies
Drosophila culture materials, including
> Medium
> Vials
> Vial plug
> Ethanol solutions:
>> 1%
>> 2%
>> 4%
>> 8%
> brushes
> index cards
> FlyNap™

Objectives
4) To determine whether there are dose-dependent effects of ethanol on the fecundity, hatch weight, or sex ratio of fruit flies;
5) To correlate these results with those collected from experiments with rodents or other experimental mammals.

Background

Ethanol freely crosses the placental barrier, and for that reason, the ingestion of ethanol by pregnant females is associated with a number of consequences:
1) A decrease in litter size;
2) A decrease in the average birth weight of individual offspring in a litter;
3) Severe damage to the corpus callosum and structures in the cerebral cortex.

Among humans, the result of these consequences is that children born to alcoholic or binging mothers present a syndrome of physical, psychological and emotional characteristics which are called **fetal alcohol syndrome**, characterized by physical retardation, poor learning and coping skills, and an inability to learn from mistakes. Studies of the brains of pups born to dams exposed to alcohol during their pregnancy have shown histological evidence of considerable damage to layers of the cerebral cortex, including a reduction in the number of neurons and a disruption in the orientation of surviving cells.

Although fruit flies do not develop *in utero*, they provide a useful model to demonstrate certain features of the consequences of early alcoholic exposure, particularly a reduction in fecundity and a reduction in the weight of newly hatched adults.

In this study, you will set up a series of vials with increasing concentrations of alcohol in order to determine whether the changes in fecundity and in hatch weight are dose-dependent.

A brief description of the biology of *Drosophila melanogaster*

For the past one hundred years, *Drosophila melanogaster*, the fruit fly, has been a model organism for studies in Mendelian and molecular genetics. Thomas Hunt Morgan documented alterations in the sequence of bands on the polytene chromosomes of *Drosophila*, and correlated those alterations with physical mutations caused by radiation. In the ensuing years, thousands of mutants strains have been isolated, and their genomes have been sequenced.

The requirements for maintaining stocks of fruit flies are simple -- a mixture of potato flakes and water in a small polypropylene vial, covered with a plug that allows the diffusion of air. At room temperature, the life cycle is completed within 2 weeks, and adults survive for approximately one month (Figure 11.1).

Adult males and females are relatively easy to distinguish under a dissecting microscope. Adults can be anesthetized with carbon dioxide or FlyNap™, and then gently moved around on an index card with a small artist's paintbrush. Males are smaller than females, have sex combs on their front legs, and have a pointed abdomen that is darkly pigmented. Females lack sex combs, and have a more rounded abdomen which is lightly pigmented (Figure 11.2).

Procedure

Setting up the fruit fly bottles
1) Work in groups of 4. Obtain 5 polypropylene vials, and label them "0%", "1%", "2%", "4%", and "8%", corresponding to the concentrations of ethanol that you will be using to prepare the medium;
2) Use the small beaker provided in the fruit fly food to dispense 12 ml of solid food, and then pour 12 ml of the appropriate liquid into each of the 5 vials. Allow each vial to stand for 5 minutes;
3) Place all the vials on their sides,
4) Anesthetize the stock fruit flies according to the instructions provided by your instructor. Place the anesthetized stock flies on an index card, and separate the males and females with an artist's paintbrush;
5) Gently place 2 males and 3 females on the wall of the vial and close the vial with a foam plug. Make sure that the fruit flies are neither caught in the fruit fly medium or crushed by the foam plug;
6) Leave the vials on their sides for at least one hour (up to overnight), until the fruit flies have regained consciousness and are capable of flying within the vial. Allow these flies to stay in the vials for one week so that they can mate and deposit eggs.

Counting and weighing the newly hatched adult flies
Within two weeks of setting up the vials, the next generation of flies will hatch. Two weeks after they've started hatching:
1) Obtain 10 vials and label them:
   a) 0% males;
   b) 0% females;
   c) 1% males;
   d) 1% females;
   e) 2% males;
   f) 2% females;
   g) 4% males;
   h) 4% females;
   i) 8% males;
   j) 8% females

2) Anesthetize the flies in each of the bottles, drop them onto an index card, and sort them according to gender. Count them, and then place them into the appropriate labeled vial according to ethanol concentration and gender. When you have completed counting the flies, place the vials into the freezer;

3) Use an analytical balance to weigh the males and females, and then calculate an average mass for males and females at each concentration of ethanol;

4) Collect the data from all groups, and then enter the data in Table 11.1;

5) Transfer the data pertaining to total number of flies hatched at each concentration and average masses of males and females to Table 11.2. Calculate group averages at each concentration. Construct histograms to show the average mass at all concentrations for males and females, and then conduct an analysis of variance (ANOVA) to determine whether the differences are significant;

6) Conduct an analysis of variance (ANOVA) on the numbers of flies hatching at each concentration to determine whether the differences are significant.

Figure 11.1. Life cycle of *Drosophila melanogaster*, the fruit fly. From: http://www.mtholyoke.edu/~cwoodard/biol210/flylifcyc.gif

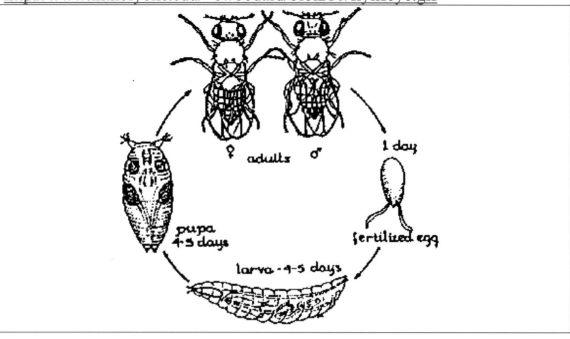

Figure 11.2. *Drosophila melanogaster* adults. Note that the female is slightly larger, has a lightly pigmented abdomen, and is lacking sex combs in the front pair of legs. The male is shorter, has a darkly pigmented abdomen, and has black sex combs in the front pair of legs (which are not conspicuous in this diagram.) These characteristics *must* be seen under a dissecting microscope for definitive identification.

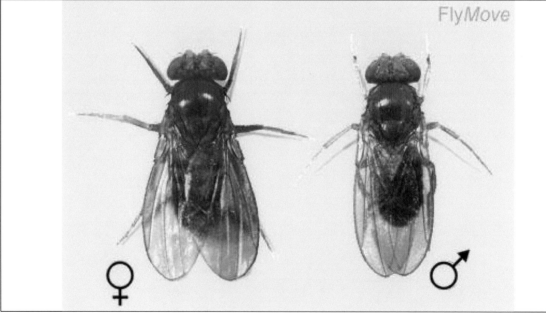

Table 11.1. Preliminary class data for experiment on the effects of ethanol on the fruit fly, *Drosophila melanogaster*. (Mass is in grams.)				
	Group 1 data	Group 2 data	Group 3 data	Group 4 data
Number of adult males hatching at 0%				
Number of adult females hatching at 0%				
Total number of adult flies hatching at 0%				
Total mass of adult males hatched at 0%				
Total mass of adult females hatched at 0%				
Average mass of males hatched at 0%				
Average mass of females hatched at 0%				
Number of adult males hatching at 1%				
Number of adult females hatching at 1%				
Total number of adult flies hatching at 1%				
Total mass of adult males hatched at 1%				
Total mass of adult females hatched at 1%				
Average mass of males hatched at 1%				
Average mass of females hatched at 1%				
Number of adult males hatching at 2%				
Number of adult females hatching at 2%				
Total number of adult flies hatching at 2%				
Total mass of adult males hatched at 2%				
Total mass of adult females hatched at 2%				
Average mass of males hatched at 2%				
Average mass of females hatched at 2%				
Number of adult males hatching at 4%				

Number of adult females hatching at 4%				
Total number of adult flies hatching at 4%				
Total mass of adult males hatched at 4%				
Total mass of adult females hatched at 4%				
Average mass of males hatched at 4%				
Average mass of females hatched at 4%				
Number of adult males hatching at 8%				
Number of adult females hatching at 8%				
Total number of adult flies hatching at 8%				
Total mass of adult males hatched at 8%				
Total mass of adult females hatched at 8%				
Average mass of males hatched at 8%				
Average mass of females hatched at 8%				

	Group #1	Group #2	Group #3	Group #4	Average of Groups 1 through 4
Table 11.2. Composite data for experiment on the effects of ethanol on *Drosophila melanogaster.* (Mass is in grams)					
Number of adult flies hatching at 0%					
Number of adult flies hatching at 1%					
Number of adult flies hatching at 2%					
Number of adult flies hatching at 4%					
Number of adult flies hatching at 8%					
Average mass of hatching adult males at 0%					
Average mass of hatching adult males at 1%					
Average mass of hatching adult males at 2%					
Average mass of hatching adult males at 4%					
Average mass of hatching adult males at 8%					
Average mass of hatching adult females at 0%					
Average mass of hatching adult females at 1%					
Average mass of hatching adult females at 2%					
Average mass of hatching adult females at 4%					
Average mass of hatching adult females at 8%					

Exercise 12: The excretory system and the detection of drug metabolites

(adapted from Warren Dolphin's Biological Investigations, edition 6.)

Equipment

Compound microscope
Dissecting microscope
Dissecting tools

Supplies

Human torso model
Fetal pigs
Cow kidneys - preserved
Human kidneys - models
Prepared slides
      Kidney l.s.
      Kidney c.s.
Urinalysis test kit for the detection of amphetamine, opiate, marijuana, and cocaine metabolites
Poppy seed cake

Objectives

1) To understand the gross anatomy of the excretory system and the histology of the kidney;
2) To demonstrate the ability of detecting metabolites in urine.

Background

The excretory systems of animals maintain a constant chemical state in the internal body fluids. The name *excretory system,* with its implicit emphasis on waste removal, does not suggest the three other important functions of these systems: (1) controlling water volume, (2) regulating salt concentrations, and (3) eliminating nonmetabolizable compounds absorbed from food or other ingested substances.

Excretion involves the elimination of the waste procuts from the metabolism of nitrogen-containing compounds. **Ammonia**, with a chemical formula of

NH_3, is a toxic compound at high concentrations and is produced by all animals when they metabolize amino acids and nucleotides. Depending on whether an animal inhabits an aquatic or terrestrial environment, this toxic product is handled in different ways. Aquatic organisms generally excrete the NH_3 directly into the surrounding water, often through their gills. Terrestrial animals convert it into less toxic compounds, such as **urea** in mammals or **uric acid** in insects, reptiles, and birds. These products are collected by excretory organs and periodically voided.

Mammalian kidneys function by filtering wastes and then selectively reabsorbing materials from the filtrate. The mammalian kidney consists of millions of capillary tufts, the **glomeruli.** They filter blood so that the fluid portion (that is, the portion with no blood cells or proteins) enters the tubules of the urinary system. The concentrations of water, salts, urea, sugars, amino acids, and so in, in the filtrate are very similar to those in the blood. As the filtrate passes through the tubular network of the **nephrons** toward the **collecting ducts** of the urinary system, many of these materials are reabsorbed by active and passive transport, leaving wastes, excess salts, and water. By the time the filtrate reaches the collecting ducts, it has been changed by the reabsorption of water and salts, but not waste products.

In the adult human, 180 liters of filtrate are produced each day, but the daily urine volume is only 0.6 to 2.0 liters. If we assume that the plasma volume is 3 liters, the production of 180 liters of filtrate means that all of the blood plasma must be filtered through the glomeruli 60 times in 24 hours. A volume of fluid equal to 178 to 179 liters is reabsorbed along with many salts. The reabsorption process is obviously of some magnitude.

Mammalian Excretory System

Anatomy of Fetal Pig Excretory System

⊠Use your fetal pig and the human torso model to study the urogenital system (Figure 12.1). Refer to these as you locate each of the structures indicated in boldface type. Be sure to look at a pig of the opposite sex before you leave the lab.

The kidneys are located on the dorsal wall of the abdominal cavity. Remove the **peritoneum,** the connective tissue membrane that holds them in place. Note the kidneys proximity to the **descending aorta.** The blood pressure drops very little as blood passes from the aorta into the kidneys via the **renal**

arteries. A high blood pressure is essential to force-filter the blood through the walls of the **glomerular capillaries. Renal veins** drain the kidney, returning blood to the caudal vena cava.

�҉Find the **ureter,** which originates from the medial face of the kidney. Trace it caudally to where it empties into the **urinary bladder.** In the fetal pig, part of the urinary bladder extends between the two **umbilical arteries** and continues into the umbilical cord, where it is called the **allantoic duct.** After birth, the duct atrophies, becoming nonfunctional.

The **urethra** proceeds caudally from the bladder to its opening; its location depends on the sex of your pig. To observe the urethra, cut through the cartilage and muscles of the **pelvic girdle** on the medial line and press the hind legs down against the dissecting tray.

�҉Remove one kidney. Make a longitudinal cut through the kidney with a sharp razor blade in a plane parallel with the front and back of the kidney. Study the cut surface with a dissecting microscope. The concave side of the kidney is the **hilum** and the outer, convex surface, the **cortex.** The **medulla** is the compact tissue containing collecting ducts and blood vessels.

Follow the ureter into the kidney. It expands into a structure called the **renal pelvis,** which subdivides into funnel shaped **calyxes.** Urine is produced by the thousands of filtration-reabsorption units called **nephrons.** It flows through collecting ducts to the calyxes and passes by way of the ureter to the bladder for storage.

Though they are not involved in excretion, note the **adrenal glands,** which are located cranially and medially to the kidneys in the abdominal cavity. Adrenaline and noradrenaline, hormones produced by these glands, regulate a number of body functions, including heart rate, blood sugar levels, and arteriole constriction.

Histology of the kidney

�҉Obtain a prepared slide of a section from the cortex of the kidney and observe it under a 10x objective.

In the cortex are tufts of capillaries each called the **glomerulus** surrounded by a cuplike **Bowman's capsule** that function as a filter unit. You will have to search on the slide to find these structures because the plane of the slice does not pass directly through each nephron.

Afferent arterioles entering the glomeruli are large, while efferent arterioles exiting the glomeruli are small. This causes a high filtration pressure in the glomerulus. The Bowman's capsule is a funnel-like device that surrounds and encloses the glomerulus.

Blood from the renal arteries enters the glomeruli at high pressure and is filtered through the capillary walls and into the Bowman's capsule. Blood cells and proteins remain in the capillaries along with some liquid. Water and dissolved, low molecular-weight substances, such as sugars, amino acids, urea, and salts, pass into the Bowman's capsule, which is the start of the **nephron.**

Nephron structure is difficult to discern in a prepared slide, since the plane of the section rarely coincides with the plane of the nephron. Figure 12.2 shows the anatomy of the nephron. As the filtrate passes through the tubules and the **loop of Henle,** water, salts, and nutrients are reabsorbed and pass into the capillaries surrounding the tubules. These absorption processes are controlled by hormones. Some water, urea, and other waste products are not reabsorbed and consequently flow through the collecting ducts to become urine.

Figure 12.1. Gross anatomy of the excretory system.

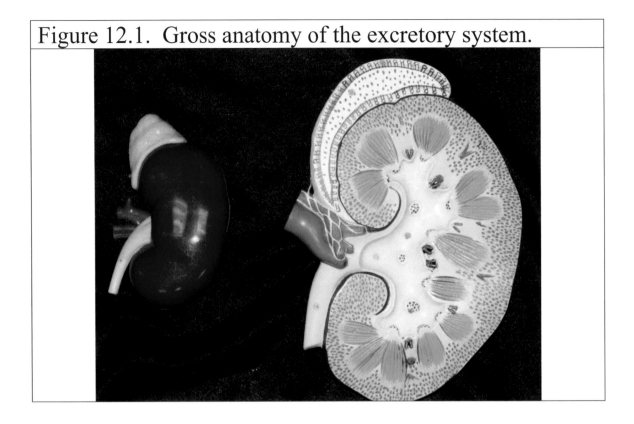

Figure 12.2. Cross-section of multiple nephron tubules. Glomerulus is at the center.

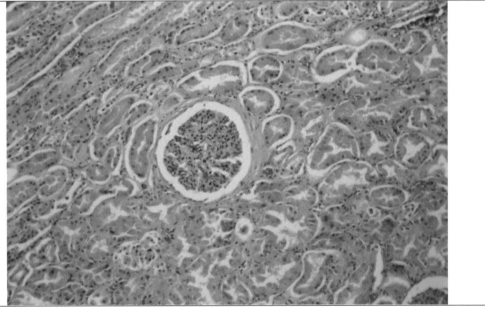

Detection of drug metabolites in urine

Background

As a drug is ingested, its blood levels will gradually as it is absorbed into the bloodstream and is distributed throughout the body. Blood levels gradually decline, in a manner shown in Figure 12.3, which would be typical of a timed, slow release drug.

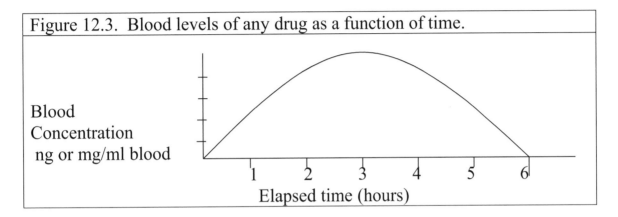

Figure 12.3. Blood levels of any drug as a function of time.

Blood Concentration ng or mg/ml blood

Elapsed time (hours)

The removal of the drug would be via one of three pathways out of the body:

1) It may be excreted in urine as the original compound without any chemical transformation;

2) It may be converted into another compound which is a characteristic or signature metabolite of that drug. Typically, metabolites are more water-soluble than the parent compound, and will be excreted in the urine;

3) The metabolite may be accumulated in bile, and then eliminated in the feces.

The appearance and persistence of a drug metabolite can be used to determine recent drug use. The drugs which are of greatest concern to employers are the following:

1) Ethanol
2) Amphetamines
3) Cocaine
4) Heroin
5) Marijuana

Of these five, only ethanol fails to produce a characteristic or signature metabolite which can be detected in urine. (Ethanol will be broken down into either acetic acid, or carbon dioxide and water, none of which would be clear indicators of ethanol use.)

The tests for the other 4 drugs (amphetamines, cocaine, heroin, and marijuana) are very sensitive, and technology has been developed to develop the presence of metabolites down to nanograms per milliliter of urine. In fact, they are so sensitive that there is occasional cross-reactivity with other compounds which are not psychoactive.

Opium is an extract of the opium poppy, and codeine, morphine, and heroin are derived from opium. Someone who is uses Tylenol-3 (Tylenol™ with codeine) or a cough medicine with codeine will produce a positive heroin test, because the metabolites of codeine, morphine, and heroin are similar enough. Even the ingestion of poppy seeds, from either rye bread or poppy seed cake, may result in a positive heroin test.

Here is the protocol for the use of the MOP Opiate Test Kit:

MOP

One Step
Opiate Test Device
Package Insert

A rapid, one step test for the qualitative detection of Opiates in human urine.

For healthcare professionals including professionals at point of care sites.

For in vitro diagnostic use only.

INTENDED USE

The MOP One Step Opiate Test Device is a lateral flow chromatographic immunoassay for the detection of opiates in urine at a cut-off concentration of 300 ng/mL. This test will detect other opiates, please refer to analytical specificity table in this package insert. **This assay provides only a preliminary analytical test result. A more specific alternate chemical method must be used in order to obtain a confirmed analytical result. Gas chromatography/mass spectrometry (GC/MS) is the preferred confirmatory method. Clinical consideration and professional judgment should be applied to any drug of abuse test result, particularly when preliminary positive results are used.**

SUMMARY

Opiate refers to any drug that is derived from the opium poppy, including the natural products, morphine and codeine, and the semi-synthetic drugs such as heroin. Opioid is more general, referring to any drug that acts on the opioid receptor.

Opioid analgesics comprise a large group of substance which control pain by depressing the central nervous system. Large doses of morphine can produce higher tolerance level and physiological dependency in users, and may lead to substance abuse. Morphine is excreted unmetabolized, and is also the major metabolic product of codeine and heroin. Morphine is detectable in the urine for several days after an opiate dose. [1]

The MOP One Step Opiate Test Device is a rapid urine screening test that can be performed without the use of an instrument. The test utilizes a monoclonal antibody to selectively detect elevated levels of opiates in urine. The MOP One Step Opiate Test Device yields a positive result when the concentration of opiates in urine exceeds the cutoff level.

PRINCIPLE

The MOP One Step Opiate Test Device is an immunoassay based on the principle of competitive binding. Drugs which may be present in the urine specimen compete against the drug conjugate for binding sites on the antibody.

During testing, a urine specimen migrates upward by capillary action. Opiates, if present in the urine specimen below the cutoff level, will not saturate the binding sites of the antibody in the test device. The morphine conjugate will be captured by antibody and a visible colored line will show up in the test line region. The colored line will not form in the test line region if the opiates level exceeds the cutoff concentration because it will saturate all the binding sites of anti-morphine antibodies.

A drug-positive urine specimen will not generate a colored line in the test line region because of drug competition, while a drug-negative urine specimen will generate a line in the test line region because of the absence of drug competition.

To serve as a procedural control, a colored line will always appear at the control line region, indicating that proper volume of specimen has been added and membrane wicking has occurred.

REAGENTS

The test device contains monoclonal anti-morphine antibody-coupled particles and morphine-protein conjugates. A goat antibody is employed in the control line system.

PRECAUTIONS

- For healthcare professionals including professionals at point of care sites.
- For professional *in vitro* diagnostic use only. Do not use after the expiration date The test device should remain in the sealed pouch until use.
- All specimens should be considered potentially hazardous and handled in the same manner as an infectious agent.
- The test device should be discarded according to federal, state and local regulations.

STORAGE AND STABILITY

Store as packaged in the sealed pouch at 2-30°C. The test device is stable through the expiration date printed on the sealed pouch. The test device must remain in the sealed pouch until use. DO NOT FREEZE. Do not use beyond the expiration date.

SPECIMEN COLLECTION AND PREPARATION

Urine Assay

The urine specimen must be collected in a clean and dry container. Urine collected at any time of the day may be used. Urine specimens exhibiting visible particles should be centrifuged, filtered, or allowed to settle to obtain a clear specimen for testing.

Specimen Storage

Urine specimens may be stored at 2-8°C for up to 48 hours prior to testing. For long-term storage, specimens may be frozen and stored below -20°C. Frozen specimens should be thawed and mixed before testing.

MATERIALS

Materials Provided

- Test devices
- Disposable specimen droppers
- Package insert

Materials Required But Not Provided

- Specimen collection container
- Timer
- External controls

DIRECTIONS FOR USE

Allow the test device, urine specimen and/or controls to reach room temperature (15-30°C) prior to testing.

1. Bring the pouch to room temperature before opening it. Remove the test device from the sealed pouch and use it as soon as possible.
2. Place the test device on a clean and level surface. Hold the dropper vertically and transfer 3 full drops of urine (approx. 100μl) to the specimen well (S) of the test device, and then start the timer. Avoid trapping air bubbles in the specimen well (S). See the illustration below.
3. Wait for the colored line(s) to appear. The result should be read at 5 minutes. Do not interpret the result after 10 minutes.

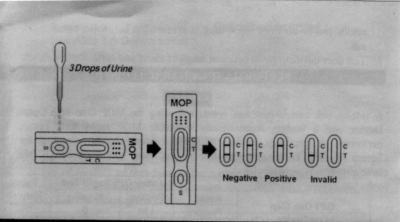

INTERPRETATION OF RESULTS

(Please refer to illustration above)

NEGATIVE:* Two lines appear. One colored line should be in the control region (C), and another colored line should be in the test region (T). This negative result indicates that the morphine concentration is below the detectable level (300 ng/mL).
* NOTE: The shade of the color in the test region (T) may vary, but it should be considered negative whenever there is even a faint line.

POSITIVE: One colored line appears in the control region (C). No line appears in the test region (T). This positive result indicates that the morphine concentration exceeds the detectable level (300 ng/mL).

INVALID: Control line fails to appear. Insufficient specimen volume or incorrect procedural techniques are the most likely reasons for control line failure. Review the procedure and repeat the test with a new test device. If the problem persists, discontinue using the test kit immediately and contact your local distributor.

QUALITY CONTROL

A procedural control is included in the test. A colored line appearing in the control region (C) is considered an internal procedural control. It confirms sufficient specimen volume, adequate membrane wicking and correct procedural technique.

Control standards are not supplied with this kit; however it is recommended that positive and negative controls be tested as good laboratory testing practice to confirm the test procedure and to verify proper test performance.

LIMITATIONS

1. The MOP One Step Opiate Test Device provides only a qualitative, preliminary analytical result. A secondary analytical method must be used to obtain a confirmed result. Gas chromatography/mass spectrometry (GC/MS) is the preferred confirmatory method. [2,3]

2. It is possible that technical or procedural errors, as well as other interfering substances in the urine specimen may cause erroneous results.

3. Adulterants, such as bleach and/or alum, in urine specimens may produce erroneous results regardless of the analytical method used. If adulteration is suspected, the test should be repeated with another urine specimen.

4. Certain medications containing opiate derivatives may produce a positive result. Additionally, foods and tea containing poppy products (the origin of the opiate) may also produce a positive result.

5. A Positive Result indicates presence of the drug or its metabolites but does not indicate level of intoxication, administration route or concentration in urine.

6. A Negative Result may not necessarily indicate drug-free urine. Negative

results can be obtained when drug is present but below the cutoff level of the test.

7. Test does not distinguish between drugs of abuse and certain medications.

PERFORMANCE CHARACTERISTICS

Accuracy

A side-by-side comparison was conducted using the MOP One Step Opiate Test Device and a leading commercially available MOP rapid test. Testing was performed on specimens previously collected from subjects presenting for Drug Screen Testing. Presumptive positive results were confirmed by GC/MS. The following results were tabulated:

Method		Other OPI Rapid Test		Total Results
	Results	Positive	Negative	
OPI One Step Test Device	Positive	150	0	150
	Negative	0	150	150
Total Results		150	150	300
% Agreement with this commercial kit		>99%	>99%	>99%

When compared to GC/MS at the cut-off of 300 ng/mL, the following results were tabulated:

Method		GC/MS		Total Results
	Results	Positive	Negative	
OPI One Step Test Device	Positive	141	9	150
	Negative	0	150	150
Total Results		141	159	300
% Agreement with GC/MS Analysis		>99%	94%	97%

Eighty (80) of these samples were also run using the MOP One Step Opiate Test Device by an untrained operator at a different site. Based on GC/MS data, the operator obtained a statistically similar Positive Agreement, Negative Agreement and Overall Agreement rate as the laboratory personnel.

Analytical Sensitivity

A drug-free urine pool was spiked with Morphine at the following concentrations: 0 ng/mL, 150 ng/mL, 225 ng/mL, 300 ng/mL, 375 ng/mL and 450 ng/mL. The result demonstrates >99% accuracy at 50% above and 50% below the cut-off concentration. The data are summarized below:

Morphine Concentration (ng/mL)	Percent of Cutoff	n	Visual Result	
			Negative	Positive
0	0%	30	30	0
150	-50%	30	30	0
225	-25%	30	28	2
300	Cutoff	30	20	10
375	+25%	30	3	27
450	+50%	30	0	30

Specificity

The following table lists compounds that are positively detected in urine by the MOP One Step Opiate Test Device at 5 minutes.

Compound	Concentration (ng/mL)
Codeine	300
Ethylmorphine	6,250
Hydrocodone	50,000
Hydromorphone	3,125
Levophanol	1,500
6-Monoacetylmorphine	400
Morphine	300

Compound	Concentration (ng/mL)
Morphine 3-β-D-glucuronide	1,000
Norcodeine	6,250
Normorphone	100,000
Oxycodone	30,000
Oxymorphone	100,000
Procaine	15,000
Thebaine	6,250

Precision

A study was conducted at 3 physician's offices by untrained operators using 3 different lots of product to demonstrate the within run, between run and between operator precision. An identical panel of coded specimens blind labeled and tested at each site. The results are given below:

Morphine conc. (ng/mL)	n	Site A		Site B		Site C	
		-	+	-	+	-	+
0	45	15	0	15	0	15	0
150	45	15	0	15	0	15	0
225	45	12	3	11	4	13	2
375	45	4	11	0	15	7	8
450	45	1	14	2	13	0	15

Effect of Urinary Specific Gravity

Fifteen (15) urine samples of normal, high, and low specific gravity ranges were spiked with 150 ng/ml and 450 ng/ml of Morphine respectively. The MOP One Step Opiate Test Device was tested in duplicate using the fifteen neat and spiked urine samples. The results demonstrate that varying ranges of urinary specific gravity does not affect the test results.

Effect of the Urinary pH

The pH of an aliquoted negative urine pool was adjusted to a pH range of 5 to 9 in 1 pH unit increments and spiked with Morphine to 150 ng/ml and 450 ng/ml. The spiked, pH-adjusted urine was tested with the MOP One Step Opiate Test Device in duplicate and interpreted according to the package insert. The results demonstrate that varying ranges of pH does not interfere with the performance of the test.

Cross-Reactivity

A study was conducted to determine the cross-reactivity of the test with compounds in either drug-free urine or morphine positive urine. The following compounds show no cross-reactivity when tested with the MOP One Step Opiate Test Device at a concentration of 100 μg/mL.

Non Cross-Reacting Compounds

4-Acetamidophenol	Erythromycin	Oxymetazoline
Acetophenetidin	β-Estradiol	Papaverine
N-Acetylprocainamide	Estrone-3-sulfate	Penicillin-G
Acetylsalicylic acid	Ethyl-p-aminobenzoate	Pentazocine
Aminopyrine	Fenoprofen	Pentobarbital
Amitryptyline	Furosemide	Perphenazine
Amobarbital	Gentisic acid	Phencyclidine
Amoxicillin	Hemoglobin	Phenelzine
	Hydralazine	Phenobarbital
L-Ascorbic acid	Hydrochlorothiazide	Phentermine
D,L-Amphetamine	Hydrocortisone	L-Phenylephrine
Apomorphine	Ọ-Hydroxyhippuric acid	β-Phenylethylamine
Aspartame	p-Hydroxy-	Phenylpropanolamine

Atropine	methamphetamine	Prednisone
Benzilic acid	3-Hydroxytyramine	D,L-Propanolol
Benzoic acid	Ibuprofen	D-Propoxyphene
Benzoylecgonine	Imipramine	D-Pseudoephedrine
Benzphetamine	Iproniazid	Quinidine
Bilirubin	(±) Isoproterenol	Quinine
(±)-Brompheniramine	Isoxsuprine	Ranitidine
Caffeine	Ketamine	Salicylic acid
Cannabidiol	Ketoprofen	Secobarbital
Chloralhydrate	Labetalol	Serotonin (5-
Chloramphenicol	Loperamide	Hydroxytyramine)
Chlordiazepoxide	Maprotiline	Sulfamethazine
Chlorothiazide	Meperidine	Sulindac
(±) Chlorpheniramine	Meprobamate	Temazepam
Chlorpromazine	Methadone	Tetracycline
Chlorquine	Methoxyphenamine	Tetrahydrocortisone, 3
Cholesterol	(+) 3,4-Methylenedioxy-	Acetate
Clomipramine	amphetamine	Tetrahydrocortisone 3 (β-D
Clonidine	(+) 3,4-Methylenedioxy-	glucuronide)
Cocaine hydrochloride	methamphetamine	Tetrahydrozoline
Cortisone	Nalidixic acid	Thiamine
(-) Cotinine		Thioridazine
Creatinine	Naloxone	D, L-Tyrosine
Deoxycorticosterone	Naltrexone	Tolbutamide
Dextromethorphan	Naproxen	Triamterene
Diazepam	Niacinamide	Trifluoperazine
Diclofenac	Nifedipine	Trimethoprim
Diflunisal	Norethindrone	Trimipramine
Digoxin	D-Norpropoxyphene	Tryptamine
Diphenhydramine	Noscapine	D, L-Tryptophan
Doxylamine	D,L-Octopamine	Tyramine
Ecgonine hydrochloride	Oxalic acid	Uric acid
Ecgonine methylester	Oxazepam	Verapamil
(-) Ψ Ephedrine	Oxolinic acid	Zomepirac

BIBLIOGRAPHY

1. Tietz NW. *Textbook of Clinical Chemistry.* W.B. Saunders Company. 1986; 1735

2. Baselt RC. Disposition of Toxic Drugs and Chemicals in Man. 2nd Ed. Biomedical Publ., Davis, CA. 1982; 488

3. Hawks RL, CN Chiang. *Urine Testing for Drugs of Abuse.* National Institute for Drug Abuse (NIDA), Research Monograph 73, 1986

Printed in China

DN: 1150058902
Eff. Date: 2005-03-10

Here is what a negative test looks like:

Here is what a positive test looks like:

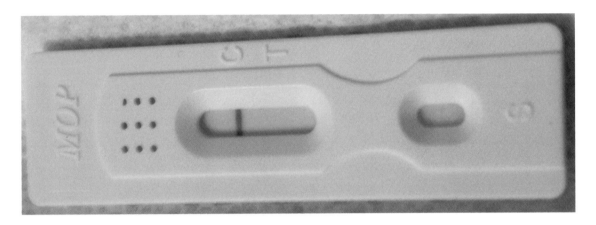

The purpose of this experiment is to demonstrate the presence of opium metabolites. Volunteers will provide a urine sample prior to ingestion of either a medical dose of Tylenol-3 or a serving of poppy seed cake to establish a baseline demonstration of a negative test. One hour following the ingestion of Tylenol-3 or poppy seed cake, volunteers will provide a second urine sample for testing.

✠Protocol
1) Volunteers should fill out a release form and submit it prior to the beginning of the experiment;
2) Volunteers will provide a baseline urine sample for analysis prior to ingestion for testing. Follow the directions in determining the presence of cocaine, amphetamine, marijuana and heroin. **All results will remain confidential.** If the baseline urine sample is positive for heroin, the student should not continue with the rest of the experiment;

3) The volunteer will then ingest one medical dose of Tylenol-3™ or a serving of poppy seed cake. The recipe for the poppy seed cake used in this experiment is described in Table 12.1;

4) One hour after ingestion of the medication or cake, the volunteer will provide a second urine sample for analysis. If there are enough test kits available, the student will then test a third urine sample 24 hours after ingestion. Record all results in Table 12.2.

✂ Exercise: Detection of opiate metabolites following ingestion pastries containing poppy seed

Release executed on this _____ day of _____,
_____(Year).

As a lawful consideration for being permitted by Christian Brothers University to participate in the below-described activity, I, the undersigned, hereby agree that I, my family, heirs, distributees, guardians, legal representatives and assigns:

1. Will not make a claim against, sue, attach the property of, or prosecute Christian Brothers University or any of its employees, agents, or instructors, (hereafter "CBU"), for any injury or damage that may occur during my voluntary participation in such activity, even though liability may arise out of the negligence or carelessness on the part of the persons or entities mentioned above.

2. Will save and hold harmless CBU from any claims by me, or my family, heirs, distributees, guardians, legal representatives or assigns arising out of my voluntary participation in such activity.

I hereby personally assume all risks of loss, damage, or injury that I may sustain in connection therewith, whether those risks are foreseen or unforeseen, and whether by negligence or not.

I further release all agents and employees from any claim whatsoever on account of first aid, treatment or service rendered me during my participation in such activity.

I further state that I am of lawful age and legally competent to sign this release. This release contains the entire agreement between the parties hereto and the terms of this release are contractual and not a mere recital.

I HAVE CAREFULLY READ THIS AGREEMENT AND FULLY UNDERSTAND ITS CONTENTS. I AM AWARE THAT THIS IS A RELEASE OF LIABILITY AND A CONTRACT BETWEEN MYSELF AND CHRISTIAN BROTHERS UNIVERSITY AND/OR ITS AGENTS AND EMPLOYEES, AND I VOLUNTARILY SIGN THIS WAIVER AND RELEASE.

DESCRIPTION OF ACTIVITY:
1) Volunteers will submit a baseline urine sample for analysis prior to ingestion of either the medication. **All results will remain confidential;**
2) Volunteers will then ingest 1 medical dose of Tylenol-3™ or a serving of poppy seed cake;
3) One hour after ingestion, volunteers will then submit a second urine sample for analysis;
4) If sufficient test kits are available, volunteers will test their urine 24 hours after ingestion of the medication or cake.

RISKS:
If the volunteer ingests the Tylenol-3, (s)he may experience slight drowsiness. Allergic reactions or side effects are not expected at this low dose. If the volunteer ingests the poppy seed cake, no allergic reactions or side effects are anticipated.

Signature --

Table 12.1. Recipe for poppy seed cake used in this experiment. This recipe is a specialty of the Crystal lake Lodge Bed and Breakfast of Mentone, Alabama, and appeared at http://www.virtualcities.com/ons/al/n/alnb2015.htm

Ingredients
    3 cups flour
    2-1/2 cups sugar
    1-1/2 teaspoons salt
    1-1/2 teaspoons baking soda
    3 eggs
    1-1/2 cups milk
    3/4 cup oil
    2-1/2 Tablespoons poppy seeds
    1-1/2 teaspoons vanilla
    1-1/2 teaspoons almond extract
 	
 Preheat oven to 325 degrees. Mix all ingredients together and pour into a greased and floured bundt pan or 2 loaf pans. Bake for 1 hour or until toothpick inserted in middle comes out clean.

Table 12.2. Results of detection of opiate metabolites experiment. (Baseline urinalysis must be negative for volunteers to participate.)

Volunteer initials	Volunteer gender	What did the volunteer ingest?	Results of baseline urinalysis test for opiates	Results of urinalysis test for opiates that evening	Results of urinalysis test for opiates the next morning
			-		
			-		
			-		
			-		

Exercise 13. OK, OK, what about *chocolate?* (Gender differences in responding to a Godiva™ chocolate tasting session and in reporting favorite flavors)

Equipment
VCR (for viewing a video on the production of Godiva™ chocolates)
Computer with Microsoft Excel or other data analysis software

Supplies
Guest book
Information booklets about Godiva™ chocolates
Video camcorder
Free samples of chocolate (at least 3 different flavors) (It helps to do this lab right after Easter, when Godiva will have a surplus of chocolate Easter eggs.)

Objectives
To determine whether there are gender-specific differences in behavior pertaining to:
1) the manner in which females and males respond to an invitation to attend a chocolate tasting session;
2) the number of chocolate pieces sampled during the chocolate tasting session;
3) the time spent in the chocolate tasting session; and
4) the favorite flavors reported by attendees

Background

For thousands of years, the beans of the cacao plant, *Theobroma cacao*, have been used to roasted and ground to prepare a beverage. In Aztec times, the bitter drink was reserved for the priests and royalty, and was consumed with hot peppers. Following the settling of the Americas by the Spaniards, sugar, and then milk were added to the cocoa liquor to produce a much sweeter concoction that became extremely popular in Europe. Years later, paraffin was added to the chocolate liquor to solidify it at room temperature, thus producing the chocolate bar.

Many of the compounds in the cacao extract are psychoactive. The following text comes from http://www.exploratorium.edu/exploring/exploring_chocolate/choc_8.html , written by Jim Spadaccini:

One of the most pleasant effects of eating chocolate is the "good feeling" that many people experience after indulging. Chocolate contains more than 300 known chemicals. Scientists have been working on isolating specific chemicals and chemical combinations which may explain some of the pleasurable effects of consuming chocolate.

Caffeine is the most well known of these chemical ingredients, and while it's present in chocolate, it can only be found in small quantities. Theobromine, a weak stimulant, is also present, in slightly higher amounts. The combination of these two chemicals (and possibly others) may provide the "lift" that chocolate eaters experience.

Phenylethylamine is also found in chocolate. It's related to amphetamines, which are strong stimulants. All of these stimulants increase the activity of neurotransmitters (brain chemicals) in parts of the brain that control our ability to pay attention and stay alert.

While stimulants contribute to a temporary sense of well-being. There are other chemicals and other theories as to why chocolate makes us feel good. Perhaps the most controversial findings come from researchers at the Neurosciences Institute in San Diego, California. They believe that "chocolate contains pharmacologically active substances that have the same effect on the brain as marijuana, and that these chemicals may be responsible for certain drug-induced psychoses associated with chocolate craving." We talked to Emmanuelle diTomaso, who worked on the original study in San Diego (she's now a researcher at Harvard), and to Daniel Piomelli, who heads the project and continues to do research at the Neurosciences Institute.

How does this work? Brain cells have a receptor for THC (tetrahydrocannabinol), which is the active ingredient in marijuana. A receptor is a structure on the surface of a cell that can lock onto certain molecules, making it possible to carry a signal through the cell wall. (diTomaso described it as a "lock-and-key" system.) "The active compound," she told me, "will lock itself to the protein on the membrane of the cell, and that triggers a reaction inside the cell." In the case of THC, that chemical reaction is what would make someone feel "high."

THC, however, is not found in chocolate. Instead, another chemical, a neurotransmitter called anandamide, has been isolated in chocolate. Interestingly, anandamide is also produced naturally in the brain. Both diTomaso and Piomelli went to great lengths to explain that this finding doesn't mean that eating chocolate will get you high, but rather that there are compounds in chocolate that may be associated with the good feeling that chocolate consumption provides.

Still, the research results made for great newspaper headlines. In 1996, when Piomelli's first study was published and "picked up" by the press, he received a number a phone calls and visits from representatives of the major chocolate companies. "They were worried," he said, "that they would have to put a warning from the Surgeon General on their products."

Anandamide, like other neurotransmitters, is broken down quickly after it's produced. Piomelli and his team found other chemicals in chocolate which may inhibit the natural breakdown of anadamide. This means that natural anandamide (or introduced anandamide) may stick around longer, making us feel good longer, when we eat chocolate.

More research needs to be done to understand the effects of chocolate on the brain, and Piomelli's group is currently working on a new study that should be published next year. In the meantime, I'm going to be doing a few experiments of my own. Now that I know more about the captivating confection, I guess I'm going to have to start sampling all the different types and brands of chocolate at my local candy store--one by one.

A number of studies have shown gender-related differences in behavioral responses to chocolate, such as those listed in Table 13.1:

Table 13.1. Studies pertaining to gender-related differences in behavioral responses to chocolate.
Small DM, Zatorre RJ, Dagher A, et al. Changes in brain activity related to eating Chocolate: from pleasure to aversion. Brain 2001 Sep;124(Pt 9):1720-33 (ISSN: 0006-8950) Small DM; Zatorre RJ; Dagher A; Evans AC; Jones-Gotman M Neuropsychology/Cognitive Neuroscience Unit, and Northwestern Cognitive Brain Mapping Group, Northwestern University School of Medicine, Chicago, IL 60611, USA. d-small@northwestern.edu. We performed successive H(2)(15)O-PET scans on volunteers as they ate chocolate to beyond satiety. Thus, the sensory stimulus and act (eating) were held constant while the reward value of the chocolate and motivation of the subject to eat were manipulated by feeding. Non-specific effects of satiety (such as feelings of fullness and autonomic changes) were also present and probably contributed to the modulation of brain activity. After eating each piece of chocolate, subjects gave ratings of how pleasant/unpleasant the chocolate was and of how much they did or did not want another piece of chocolate. Regional cerebral blood flow was then regressed against subjects' ratings. Different groups of structures were recruited selectively depending on whether subjects were eating chocolate when they were highly motivated to eat and rated the chocolate as very pleasant [subcallosal region, caudomedial orbitofrontal cortex (OFC), insula/operculum, striatum and midbrain] or whether they ate chocolate despite being satiated (parahippocampal gyrus, caudolateral

OFC and prefrontal regions). As predicted, modulation was observed in cortical chemosensory areas, including the insula and caudomedial and caudolateral OFC, suggesting that the reward value of food is represented here. Of particular interest, the medial and lateral caudal OFC showed opposite patterns of activity. This pattern of activity indicates that there may be a functional segregation of the neural representation of reward and punishment within this region. The only brain region that was active during both positive and negative compared with neutral conditions was the posterior cingulate cortex. Therefore, these results support the hypothesis that there are two separate motivational systems: one orchestrating approach and another avoidance behaviours.

Dillinger TL, Barriga P, Escarcega S, et al. Food of the gods: cure for humanity? A Cultural history of the medicinal and ritual use of chocolate. J Nutr 2000 Aug;130(8SSuppl):2057S-72S (ISSN: 0022-3166)

Dillinger TL; Barriga P; Escarcega S; Jimenez M; Salazar Lowe D; Grivetti LE . Department of Nutrition, University of California, Davis CA 95616, USA.

The medicinal use of cacao, or chocolate, both as a primary remedy and as a vehicle to deliver other medicines, originated in the New World and diffused to Europe in the mid 1500s. These practices originated among the Olmec, Maya and Mexica (Aztec). The word cacao is derived from Olmec and the subsequent Mayan languages (kakaw); the chocolate-related term cacahuatl is Nahuatl (Aztec language), derived from Olmec/Mayan etymology. Early colonial era documents included instructions for the medicinal use of cacao. The Badianus Codex (1552) noted the use of cacao flowers to treat fatigue, whereas the Florentine Codex (1590) offered a prescription of cacao beans, maize and the herb tlacoxochitl (Calliandra anomala) to alleviate fever and panting of breath and to treat the faint of heart. Subsequent 16th to early 20th century manuscripts produced in Europe and New Spain revealed >100 medicinal uses for cacao/chocolate. Three consistent roles can be identified: 1) to treat emaciated patients to gain weight; 2) to stimulate nervous systems of apathetic, exhausted or feeble patients; and 3) to improve digestion and elimination where cacao/chocolate countered the effects of stagnant or weak stomachs, stimulated kidneys and improved bowel function. Additional medical complaints treated with chocolate/cacao have included anemia, poor appetite, mental fatigue, poor breast milk production, consumption/tuberculosis, fever, gout, kidney stones, reduced longevity and poor sexual appetite/low virility. Chocolate paste was a medium used to administer drugs and to counter the taste of bitter pharmacological additives.

In addition to cacao beans, preparations of cacao bark, oil (cacao butter), leaves and flowers have been used to treat burns, bowel dysfunction, cuts and skin irritations.

Schroeder BE, Binzak JM, Kelley AE A common profile of prefrontal cortical activation Following exposure to Nicotine- or chocolate-associated contextual cues. Neuroscience 2001;105(3):535-45 (ISSN: 0306-4522)

Schroeder BE; Binzak JM; Kelley AE Neuroscience Training Program, University of Wisconsin-Madison 53705, USA. beschroe@students.wisc.edu.

Conditioning and learning factors are likely to play key roles in the process of addiction and in relapse to drug use. In nicotine addiction, for example, contextual cues associated with smoking can be powerful determinants of craving and relapse, even after considerable periods of abstinence. Using the detection of the immediate-early gene product, Fos, we examined which regions of the brain are activated by environmental cues associated with nicotine administration, and compared this profile to the pattern induced by cues associated with a natural reward, chocolate. In the first experiment, rats were treated with either nicotine (0.4 mg/ml/kg) or saline once per day for 10 days in a test environment distinct from their home cages. In the second experiment, rats were given access to either a bowl of chocolate chips or an empty bowl in the distinct environment for 10 days. After a 4-day interval, rats were re-introduced to the environment where they previously received either nicotine treatment or chocolate access. Nicotine-associated sensory cues elicited marked and specific activation of Fos expression in prefrontal cortical and limbic regions. Moreover, exposure to cues associated with the natural reward, chocolate, induced a pattern of gene expression that showed many similarities with that elicited by drug cues, particularly in prefrontal regions. These observations support the hypothesis that addictive drugs induce long-term neuroadaptations in brain regions subserving normal learning and memory for motivationally salient stimuli.

Macht M, Roth S, Ellgring H Chocolate eating in healthy men during experimentally induced sadness and joy. Appetite 2002 Oct;39(2):147-58 (ISSN: 0195-6663)

Macht M; Roth S; Ellgring H Institute for Psychology (I), University of Wurzburg, Germany. macht@psychologie.uni-wuerzburg.de.

The study compared influences of qualitatively different emotions on eating. Motivation to eat, affective responses to chocolate and chewing of chocolate were investigated in healthy normal weight males during experimentally induced emotions. Subjects abstained from eating 2 h (n = 24) or 8 h (n = 24) before testing. They received pieces of chocolate after viewing film clips presented to induce anger, fear, sadness and joy. Motivation to eat and most affective responses to eating chocolate were higher after 8 h than after 2 h of deprivation. Sadness and joy affected motivation to eat in opposite directions: joy increased and sadness decreased appetite ($p < 0.001$). In joy, a higher tendency to eat more chocolate was reported ($p < 0.001$), and chocolate tasted more pleasant ($p < 0.001$) and was experienced as more "stimulating" than in sadness ($p < 0.01$). No effects of deprivation could be found for chewing time and number of chews. Results indicate that the quality of emotions can affect motivation to eat and affective responses to eating chocolate. Our findings on decreased eating responses to sadness in healthy males and the contradictory increased eating responses to sadness reported by others supports two types of emotion-induced changes of eating: emotion-congruent modulation of eating and eating to regulate emotions. [Copyright 2002 Elsevier Science Ltd.].

The purpose of this study is to examine possible gender-related differences in the following:

1) Responding to an open invitation to a chocolate tasting;
2) Number of chocolate pieces sampled during the chocolate tasting session;
3) Favorite flavors reported by attendees.

✠Procedure

1) ONE MONTH PRIOR TO THE SESSION: Call up local chocolatiers to find out which one(s) would be willing to donate samples. During the Spring 2003 semester, Marianne Badger, manager of the Godiva ™ Chocolatier at the Oak Grove Mall was kind enough to commit to supplying us with sample chocolates, mostly surplus chocolate Easter eggs;
2) ONE WEEK PRIOR TO THE SESSION: Schedule a date, room, and time for the chocolate tasting session, and post approved flyers on campus;
3) THE DAY PRIOR TO THE SESSION: Pick up the chocolate samples, explanatory brochures and a video;
4) THE DAY OF THE SESSION: Set up the room with the chocolate samples, descriptions, and VCR. Set up the VCR so that people can see

the screen as they walk in. Set up a guest book sign-in sheet as shown in Table 13.2.

Table 13.2. Format for guest list sign-in sheet.				
Name	Time in	How many pieces of chocolate did you sample?	What was your favorite flavor?	Time out

5) POST-SESSION ANALYSIS:
   a) Determine the number of females and males who attended the session and conduct a Chi-square goodness-of-fit test to determine whether the observed distribution differs from a 50% female-50% male distribution. Use Table 13.3 for your calculations.

Table 13.3. Calculations for a Chi-square goodness-of-fit to determine whether the observed gender distribution of visitors to the chocolate tasting session differs from a 50% female-50% male distribution.

Group	Observed Number attending session	Expected number (50% each)	Observed - Expected	$(O-E)^2$	$\dfrac{(O-E)^2}{E}$
Female					
Male					
Chi-square = $\sum (O-E)^2 / E$					
Degrees of freedom (d.f.) = # of groups -1 =					1
What is the critical value for Chi-square for $p < .05$ when d.f. = 1?					
Is the observed distribution of females: males significantly different from a 50%-female:50%-male distribution? (Yes/No)					

   b) Conduct a Student's t-test to determine whether the average number of pieces sampled by females and males is significantly different. (You can use Microsoft Excel to do this.)

c) Conduct a Student's t-test to determine whether the average amount of time (in minutes) that females and males spent at the chocolate tasting session was different. (You can use Microsoft excel to do this.)

d) Conduct a Chi-square goodness-of-fit test to determine whether the distribution of favorite flavors reported by females and males is significantly different. You can do this by designating the percentage of favorite flavors reported by females as the "expected" distribution, and using the male distribution as the "observed" number.

Appendices

Appendix I. GUIDELINES FOR LABORATORY REPORTS

1. Please DOUBLE-SPACE your laboratory report, and use margin size to 1 inch.
2. Place a cover sheet at the front of your laboratory report. The cover sheet should have your name, the report title, the course and section numbers centered on the page. PLEASE DO NOT PLACE YOUR LABORATORY REPORT IN A PLASTIC OR OTHER BINDER.

The laboratory report should include the following sections:

1) An <u>ABSTRACT</u> section, in which you describe in briefest form, the purpose, primary results and conclusions of the research report. By convention, it is 200 words or 3% of the laboratory report, whichever is LESS;

2) An <u>INTRODUCTION</u> section, in which you provide information pertaining to the problem as it is recognized and in which you discuss background information which would be pertinent to the reader. The purpose, in which you specify the questions to be addressed in THIS lab report, should be in the LAST paragraph of the introduction section;

3) A <u>MATERIALS AND METHODS</u> section, in which you discuss the organism(s) under study and the experimental protocol in "text" form. PLEASE DO NOT INCLUDE A MATERIALS LIST. If there are several parts to the experiment, each part should be described separately. In text format, briefly describe the protocol you followed in conducting the experiment. If there are several parts to the experiment, each part should be described separately;

4) A <u>RESULTS</u> section, in which you discuss the data from each part of the study in the same sequence as the parts were described in the Materials and Methods section. Use a paragraph to tell the reader what the main point is, and at the end of the sentence, refer to a specific Table or FFigure, as in the following: "Seedlings exposed to either .1% or .2% phosphate grew vigorously, but the controls did not (Figure 1)." It is essential to convert or present the data in an understandable format. <u>CHARTS OF RAW DATA ARE NEITHER NECESSARY NOR DESIRABLE!</u>;

5) A <u>DISCUSSION</u> section, in which you relate the results of your experiment to the general body of knowledge pertinent to this area of research;

6) A <u>REFERENCES</u> section, in which you list the references used for background information and/or protocol procedures, including your laboratory textbook.

CRITERIA FOR THE GRADING OF PAPERS AND EXPERIMENTAL REPORTS

The maximum grade is a 4.0 and is a composite of three grades based on spelling grammar, and content.

I. <u>Spelling</u> counts 10% of the total grade. Each different spelling or typographical error will usually result in a point deducted from the maximum. However, if one word is consistently misspelled, it will be deducted only once. Low grades in spelling can be avoided by keeping a dictionary on hand and proofreading your work before you submit it for review.

II. <u>Grammar</u> counts 20% of the total grade. Each grammar error (wrong tense, poor sentence of paragraph structure) will usually result in a point deducted from the maximum. Low grades in grammar can be avoided by proofreading your work before you submit it and by writing practice essays.

III. <u>Content</u> counts 30% of the total grade. The kinds of questions that are considered in evaluating content include the following:
1. Is your information accurate?
2. Is your discussion logical?
3. Do you adequately support your argument?
4. Do you adequately correlate and contrast your data to previous experience?
5. Do you support your conclusions with the appropriate statistical test(s)?

IV. <u>Format</u> counts 40% of the total grade. The kinds of questions that are considered in evaluating format include the following:
1. Did you transform the raw data into a more useful and appropriate format?
2. Do you follow the protocol for a laboratory report as described in the Transactions of the Tennessee Academy of Sciences?

You should write your reports as if you were submitting them to the Transactions of the Tennessee Academy of Sciences. I, in turn, will review them as if I were an editor for the journal.

Name_____

<u>Grades</u>

<u>Spelling</u>_____ x   10%   =   _____.

<u>Grammar</u>_____ x   20%   =   _____.

<u>Content</u>_____ x   30%   =   _____.

<u>Format</u>_____ x   40%   =   _____.

<u>COMPOSITE GRADE</u>_____.

Some General Guidelines for Laboratory Reports

1. **Use first-person past tense in the abstract, materials & methods & results sections, since you are describing what you <u>did</u>.**

    In other words, "we dissected the liver from *Lepomis macrochirus*" is clearer than "Livers were dissected from *Lepomis macrochirus*".

2. **Species names should be *italicized* or <u>underlined</u>.**

    For example, "We studied the excystation behavior of *Posthodiplostomum minimum*" or "We studied the excystation behavior of <u>Posthodiplostomum minimum.</u>

3. **When a species has a long name, it is acceptable to contract the genus name to one letter if you refer to it as such at the beginning of your paper.**

    For example, "We studied the excystation behavior of *Posthodiplostumum minimum* (referred in this paper as *P. minimum*)."

4. **The References Cited section should include those articles or books from which you collected information and quote it in your report. The citation in your paper should appear as (AuthorLastName, YearOfPublication).**

    For example, "*P. minimum* metacercariae become resistant to pepsin between days 26 and 44 (Eisen, 1999).

5. **Each Figure should be numbered and referred to in the text of your results section in parentheses.**

    For example, "We observed maximal movement in the well where the larvae were first exposed to acid saline with pepsin, followed by alkaline Tyrode's solution with trypsin (Figure 1).

Appendix II. Making Graphs
(From Warren Dolphin's Biological Investigations)

Graphs are used to summarize data - to show the relationship between two variables. Graphs are easier to remember than are numbers in a table and are used extensively in science. You should get in the habit of making graphs of experimental data, and you should be able to interpret graphs quickly to grasp a scientific principle.

In using this lab manual, you will be asked to make two kinds of graphs -- line graphs and histograms. **Line graphs** show the relationship between two variables, such s amount of oxygen consumed by a tadpole over an extended period of time (Figure II.1). **Histograms** are bar graphs and are usually used to represent frequency data, that is, data in which measurements are repeated and the counts are recorded, such as the values obtained when an object is weighed several times (Figure II.2.)

Line graphs
When you make line graphs, always follow these rules:
1) Decide which variable is the dependent variable and which is the independent variable. The **dependent** variable is the variable you know as a result of making experimental measurements. The **independent** variable is the information you know before you start the experiment. It does not change as a result of the dependent variable but changes independently of the other variable. In Figure II.1, time does not change as a result of oxygen consumption. Therefore, time is the independent variable and the amount of oxygen consumed (which is dependent on time) is the dependent variable;
2) Always place the independent variable on the x-axis (the horizontal one), and the dependent variable on the y-axis (the vertical one);
3) *Always label* the axes with a few words describing the variable, and *always put the units* of the variable in parentheses after the variable description (Figure II.1);
4) Choose an appropriate *scale* for the dependent and independent variables so that the highest value of each will fit on the graph paper;
5) Plot the data set (the values of y for particular values of x). Make the plotted points dark enough to be seen. If two or more data sets are to be plotted on the same coordinates, use different plotting symbols for each data set (·, x, O, ◎, □, etc.);

6) Draw *smooth curves* or *straight lines* to fit the values plotted for any one data set. Do not connect the points with short lines. A smooth curve through a set of points is a visual way of averaging out chance variability in data. Do not extrapolate beyond a data set unless you are using it as a prediction technique because you do not know from your experiment whether the relationship holds beyond the range tested;

7) Every graph should have a legend, a sentence, explaining what the graph is about.

Figure II.1. Oxygen consumption by a tadpole at two temperatures

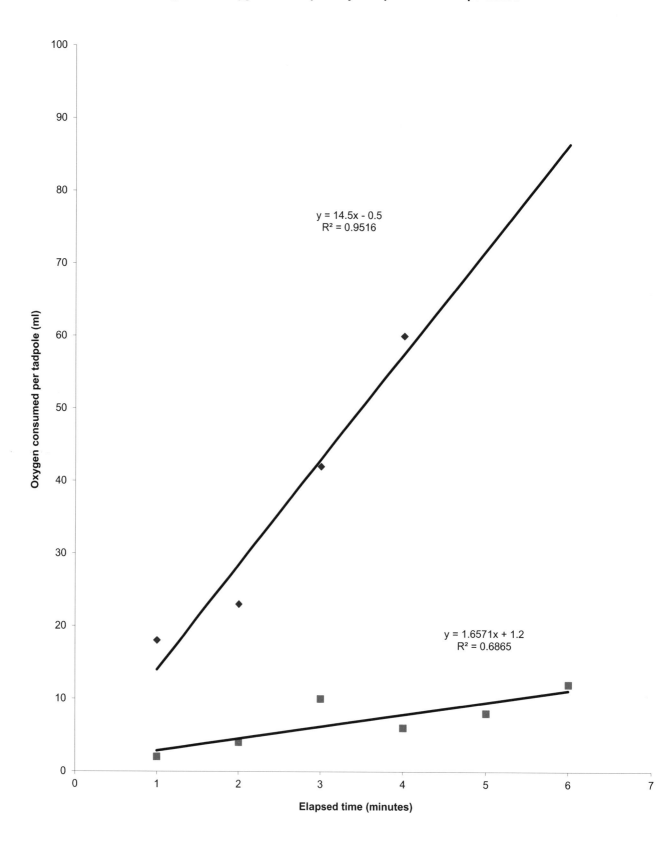

$y = 14.5x - 0.5$
$R^2 = 0.9516$

$y = 1.6571x + 1.2$
$R^2 = 0.6865$

Histograms

In making histograms, the count data are always on the y-axis. The categories in which the data fall are on the x-axis. For example, the data from which Figure II.2 was drawn are as follows:

Class Results from a Series of Weighings of the Sample (in Grams)				
61.0	60.0	60.0	59.8	58.0
61.5	61.5	61.0	60.9	58.0
59.0	60.0	60.2	61.7	60.6
59.7	59.0	60.3	63.0	60.4
62	59.0	60.7	58.5	59.0

To make the histogram, the x-axis was laid out with a range of 58 to 64 so that all values would be included. The values were then marked on the graph. After all data were plotted, bars were then drawn to show the frequencies of measurements. On bar graphs, it is good idea to show the average value across all measurements with an arrow.

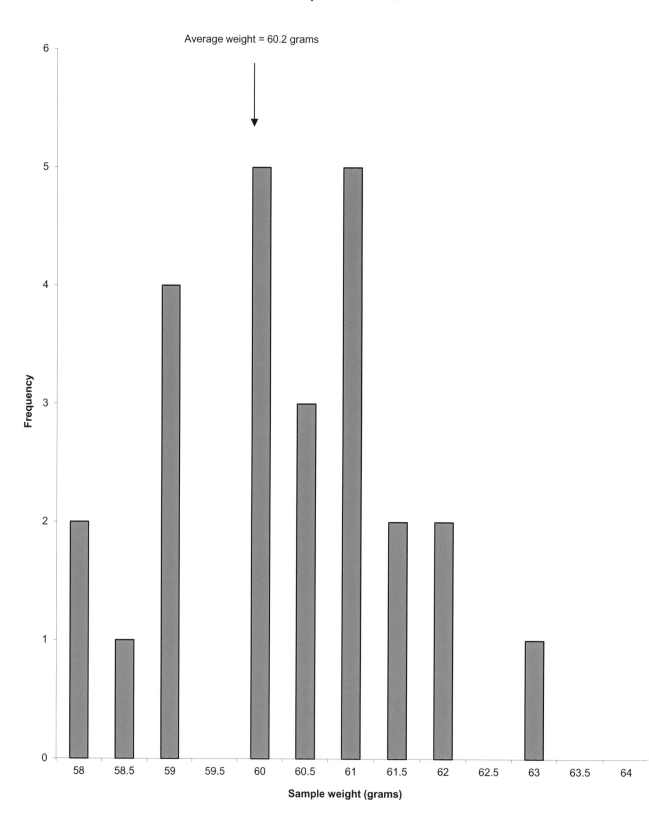

Figure II.2. Histogram of a hypothetical series of sample weights obtained by weighing the same sample several times.

Appendix III. Using Microsoft Excel for Graphing Lab Data

- Enter Excel by double clicking on the Excel icon.
  - Workbook 1 will come to the screen.

A. Plot 3 lines of best fit for the 3 sets of pH data below:

Time (sec)	Δ in Abs. @ pH3	Δ in Abs. @ pH5	Δ in Abs. @ pH7
20	0.01	0.04	0.05
40	0.01	0.06	0.05
60	0.02	0.08	0.06
80	0.02	0.11	0.07
100	0.03	0.13	0.09

1. On sheet 1 of workbook 1, type in data (not the titles) from above. Use Return key and/or arrow keys to move from square to square,
   a) Column A = Independent variable (Time data in this case).
   b) Column B = Dependent variable (Absorbance data @ 500nm, pH 3).
   c) Column C = Dependent variable (Absorbance data at 500nm, pH 5).
   d) Column D = Dependent variable (Absorbance data at 500nm, pH 7).
Hit Return

2. Click on the CHART WIZARD icon on the Menu Bar (colored mini-bar graph).

3. On the Chart Type screen, choose XY Scatter by clicking once on this icon.
   -Click once on "Next" at bottom of Chart Type screen.

4. On the Chart Source Data screen, enter the cell labels for the data by doing the following:

-Click on "Columns" under "Series in" (within the Data Range folder).

  -Now leave this folder and choose the Series folder
     a) Highlight series 1 within the Series box.
          Type "pH 3" in the Name box.

     b) Highlight series 2 and type "pH 5" in the Name box.

     c) Highlight series 3 and type "pH 7" in the Name box

5. Click on "Next" to see your graph. The labels you typed in should appear to the right of the graph.

6. On the Chart Options screen, you will get the Titles folder:
     a) Type in a descriptive title for your graph under "Title".
          -Ex) Effect of pH on Enzyme Activity
     b) Click on the Value (X) Axis box and type in this label.
          -Ex) Time (Seconds)
     c) Click on the Value (Y) Axis box and type in the label.
          -Ex) Absorbance @ 500nm

7. Click on "Next". You should now see the Chart Location screen.
     -Choose "Finish".
     -Your labeled graph will be seen with data pts but without lines of best fit.

8. For lines of best fit, pull down the CHART menu at the top of the screen.
     a) Choose "Add Trendline"
     b) On the "Type" folder, choose "pH 3" (series 1).
          -Click on "OK".

     c) Repeat steps 7a – 7c, except choose "pH 5" on the "Type" folder.
     d) Repeat steps 7a – 7c, except choose "pH 7" on the "Type" folder.

9. To get the linear regression equation for each line, go to the Chart menu:

a) Choose Trendline

b) Choose the "Options" folder

c) Click on the box next to "Display equation on chart"

B. Plot a Histogram (Bar graph) for the data above:

1. This graph will plot pH (X-axis) vs. Total Change in Abs. (Y-axis). In order to obtain data for the total change in absorb., subtract the initial absorb. value for each pH (see previous page), from the final absorb. value for each pH.

pH	Total Change in Abs.
3	0.02
5	0.09
7	0.05

2. Choose Sheet 2 at the bottom of the screen.

3. Type in data:
   a) Column A = pH data
   b) Column B = Total Change in Abs. data.

4. Highlight Column B data. (This will place the appropriate #s on the Y-axis.)

5. Click on the CHART WIZARD icon on the Tool bar.

6. On the "Chart Type" screen, choose "Column Graph".
   -Click "Next".

7. You should now see the "Chart Source Data" screen.
   a) The "Data Range" folder should come up automatically. In the "Data Range" box, you should see: =Sheet2!\$A\$1:\$B\$3
   (These are the labels for the cells containing your data. No need to alter.)

   b) Choose the "Series" folder within the "Chart Source Data" section.
      - Series 1 should be highlighted.
      - Within the "Values" box you should see: =Sheet2!\$B\$1:\$B\$3

- In the "Category X axis labels" box, type the appropriate **sheet # and cell labels for the data on the X axis. In this case,**

$$=Sheet2!\$A\$1:\$A\$3$$

(This labels the x-axis with the numbers typed into the A column.)

c) Click "Next".

8. On the "Chart Option" screen, type in your chart title, (Effect of pH **on Enzyme Activity) x-axis (pH) and y-axis (Abs.) labels.**

a) Click "Next".

b) Click "Finish".